AIRFRAME

AIRFRAME SECTION

WORKBOOK

By Dale Crane

Based on
FAA Advisory Circular 65-15A
Airframe and Powerplant Mechanics Powerplant Handbook

International Standard Book Number **0-940732-72-6**

ASA Publications, Inc.
6001 Sixth Ave. So.
Seattle, WA 98108-3307

Copyright ©1988 by ASA Publications, Inc.
All Rights Reserved
Printed in the United States of America

INTRODUCTION

The Federal Aviation Administration requires applicants for mechanic certification to pass three different types of tests: written tests, oral tests, and practical tests.

The written tests are made up of a series of multiple-choice questions taken primarily from the three FAA-published handbooks, Advisory Circulars AC65-9A, *Airframe and Powerplant Mechanics General Handbook*, AC65-12A, *Airframe and Powerplant Mechanics Powerplant Handbook*, and AC65-15A, *Airframe and Powerplant Mechanics Airframe Handbook*..

To help an applicant prepare for FAA mechanic certification, ASA offers a complete package of training materials.
- The AC65-9A, -12A, and -15A Handbooks
- Study Guides for each of these Handbooks
- Workbooks for each of these Handbooks
- Test Guides for each of the written, oral, and practical tests

The AC65 Handbooks are reprinted directly from the FAA publications and are the source of more than 80% of the questions asked on the written tests.

The Study Guides use a series of questions to guide the student through the handbooks and help them concentrate on the most important information in these books.

The Workbooks reinforce the learning process by allowing the student to write the answers to hundreds of questions on pertinent portions of the handbooks.

The ASA Test Guides contain all of the questions in the latest versions of the FAA-T-8080 written test books. Using the proven Fast-Track format, these Guides give a succinct explanation to each question so the applicant can understand the reason the choice is correct. The number for the correct answer and the reference from which the FAA took the question is provided for each question.

Included in the Test Guides are typical questions likely to be asked on the oral tests, along with their correct answer and a list of typical projects likely to be required on the practical tests.

Other ASA books of special interest to aircraft mechanics are
- ASA's Dictionary of Aeronautical Terms
- FAA Advisory Circulars AC 43.13-1A and -2A
  Acceptable Methods, Techniques, and Practices — Aircraft Inspection, Repair, and Alterations
- Study Guide for AC 43.13-1A, and -2A
- Federal Aviation Regulations for Aviation Mechanics
- Study Guide for Federal Aviation Regulations for Aircraft Mechanics

ASA is committed to providing technical help for aircraft mechanics, and it is our desire to furnish you with the materials you need. If you have any suggestions that will help us be of more service to you, please call us on our toll-free number: 1-800-426-8338.

TABLE OF CONTENTS

INTRODUCTION..iii

TABLE OF CONTENTS...iv

CHAPTER ONE
    AIRCRAFT STRUCTURES...ONE-1

CHAPTER TWO
    ASSEMBLY AND RIGGING...TWO-1

CHAPTER THREE
    AIRCRAFT FABRIC COVERING...THREE-1

CHAPTER FOUR
    AIRCRAFT PAINTING AND FINISHING...FOUR-1

CHAPTER FIVE
    AIRCRAFT STRUCTURAL REPAIRS..FIVE-1

CHAPTER SIX
    AIRCRAFT WELDING..SIX-1

CHAPTER SEVEN
    ICE AND RAIN PROTECTION..SEVEN-1

CHAPTER EIGHT
    HYDRAULIC AND PNEUMATIC POWER SYSTEMS.......................EIGHT-1

CHAPTER NINE
    LANDING GEAR SYSTEMS..NINE-1

CHAPTER TEN
    FIRE PROTECTION SYSTEMS..TEN-1

AIRFRAME

CHAPTER ELEVEN
   AIRCRAFT ELECTRICAL SYSTEMS..ELEVEN-1

CHAPTER TWELVE
   AIRCRAFT INSTRUMENT SYSTEMS...TWELVE-1

CHAPTER THIRTEEN
   COMMUNICATIONS AND NAVIGATION SYSTEMS...THIRTEEN-1

CHAPTER FOURTEEN
   CABIN ATMOSPHERE CONTROL SYSTEMS..FOURTEEN-1

CHAPTER 1

AIRCRAFT STRUCTURES

1. The five principal units of a fixed-wing aircraft airframe are

   A. _____

   B. _____

   C. _____

   D. _____

   E. _____

2. The four principal units of a helicopter are

   A. _____

   B. _____

   C. _____

   D. _____

3. The five major stresses to which an aircraft structure is subjected are

   A. _____

   B. _____

   C. _____

   D. _____

   E. _____

4. An internal force in an aircraft structure that opposes or resists deformation is called _____ (stress or strain).

5. The deformation of a material caused by forces acting on it is called _____ (stress or strain).

AIRCRAFT STRUCTURES — CHAPTER ONE

6. The force that tries to pull an object apart is called _____.

7. The stress within an object that keeps it from being crushed is called _____.

8. The stress that causes twisting is called _____.

9. The stress that tries to cause a layer of a material to slide over the layer next to it is called _____ stress.

10. A single stress resulting from tension and compression combined in a single object is called a _____ stress.

11. Two general types of fuselage construction are

   A. _____

   B. _____

12. Three classes of monocoque fuselage structures are

   A. _____

   B. _____

   C. _____

13. Primary stresses are carried in a monocoque fuselage by the _____.

14. Most modern aircraft use fuselages of the _____ type.

15. Primary bending loads in a semimonocoque fuselage are carried by the _____.

16. Longitudinal members in a semimonocoque fuselage that support the skin between the frames and longerons are called _____.

17. Vertical structural members in a semimonocoque fuselage are called frames, formers, or _____.

CHAPTER ONE   AIRCRAFT STRUCTURES

18. A connecting bracket used to join two pieces of aircraft structure is called a _____.

19. Locations along the length of an aircraft fuselage are referred to as fuselage _____.

20. The zero point from which fuselage stations are measured is called the reference _____.

21. Fuselage station 123 is located 123 _____ (what units) from the datum.

22. Reference lines used to identify a location to the right or left of the centerline of an aircraft fuselage are called _____ lines.

23. Reference lines used to identify a location above or below the horizontal reference line of an aircraft fuselage are called _____ lines.

24. An airplane wing that does not use any external braces is called a _____ wing.

25. The primary spanwise structural member in an airplane wing is called a _____.

26. Locations on an airplane wing are measured from wing station _____ which is located along the center of the fuselage.

27. Structural members that give an airplane wing its aerodynamic shape are called wing _____.

28. A wooden wing spar made up of several thin strips of wood glued together so their grain runs in the same direction is called a _____ wood spar.

29. A solid metal wing spar made by forcing metal in its plastic state through a die the shape of the spar cross section is called an _____ spar.

30. The heavy top and bottom members of an I-beam-type wing spar are called spar _____.

AIRCRAFT STRUCTURES — CHAPTER ONE

31. The upright member between the spar caps is called the _____ of the spar.

32. Partial wing ribs that extend from the leading edge back to the front spar to give the leading edge a smooth shape are called nose ribs or _____ ribs.

33. Strong steel wires that extend from the front spar, inboard to the rear spar, outboard are called _____ wires.

34. Strong steel wires that extend from the rear spar, inboard to the front spar, outboard are called _____ wires.

35. Strong ribs that separate the wing spars and carry compression loads caused by the drag and anti-drag wires are called _____ ribs.

36. An airplane wing with a portion of the structure sealed to form a fuel tank is called a _____ wing.

37. Bonded honeycomb structural panels are used in many modern aircraft to provide both strength and rigidity. The core of these honeycomb panels can be made of _____ or _____ .

38. The structure that houses the powerplant on an airplane is called an engine _____ or engine _____ .

39. The correct name for the tail section of an airplane is the _____ .

40. The primary controls that rotate an airplane about its longitudinal axis are the _____ .

41. The primary control that rotates an airplane about its vertical axis is the _____ .

42. The primary controls that rotate an airplane about its lateral axis are the _____ .

43. Four types of auxiliary controls used on an airplane are

   A. _____

CHAPTER ONE   AIRCRAFT STRUCTURES

    B. _____

    C. _____

    D. _____

44. Aileron hinges are located _____ (at or behind) the aileron's leading edge.

45. Ailerons on some high-speed airplanes are assisted in being moved and held against air loads by the use of aileron _____.

46. Deflection of wing flaps increases the aerodynamic _____ as well as the drag produced by the wing.

47. Triple-slotted, trailing-edge wing flaps increase the wing _____ as well as its camber.

48. Deflection of a leading-edge wing flap increases the _____, or curvature, of the wing.

49. Spoilers are used on an airplane to increase the drag, and in some installations to provide rotational control about the _____ axis.

50. The most commonly used method of joining sheets of aluminum alloy for aircraft construction is by _____.

51. A nonstructural panel or section of skin used to smooth the airflow over a portion of an aircraft is called a _____.

CHAPTER 2

ASSEMBLY AND RIGGING

1. Putting all of the component parts of an aircraft together is called _____.

2. The final adjustment of the alignment and movement of aircraft components is called _____.

3. The rigging specifications for an aircraft may be found in the _____ for the aircraft.

4. The two major gases in our atmosphere are _____ and _____.

5. The atmosphere presses down uniformly on the surface of the earth with a standard pressure of _____ pounds per square inch.

6. The standard weight of the atmosphere surrounding the earth will support a column of mercury _____ inches high.

7. When the atmospheric pressure increases with no change in temperature, the density of the air _____ (increases or decreases).

8. When the atmospheric temperature increases with no change in pressure, the density of the air _____ (increases or decreases).

9. With the engines producing the same horsepower, an airplane can fly faster at _____ (high or low) altitude.

10. Water vapor weighs _____ (more or less) than dry air.

11. If the total energy in the air remains constant, an increase in the velocity of the air will cause a/an _____ (increase or decrease) in the pressure of the air.

ASSEMBLY AND RIGGING   CHAPTER TWO

12. The four forces that act on an aircraft in flight are

    A. _____

    B. _____

    C. _____

    D. _____

13. Motion of an object in a particular direction with relation to time is called

    _____ (speed or velocity).

14. The rate of increase in velocity of an object is called _____.

15. According to Newton's first law of motion, a body in motion will continue in motion, at the same speed and in a straight line, until it is acted on by some

    _____.

16. The change in speed of a body in motion is determined by the amount of external force applied to the body and to the _____ of the body.

17. Five things that affect the amount of aerodynamic lift produced by an airplane wing are

    A. _____

    B. _____

    C. _____

    D. _____

    E. _____

18. The acute angle formed between the chordline of an airfoil and the line of the relative airstream (relative wind) is called the angle of _____.

19. The point at which the resultant lift and the chordline of an airfoil intersect is called the center of _____.

20. In normal flight conditions, the center of pressure of an asymmetrical airfoil moves _____ (forward or backward) as the angle of attack increases.

CHAPTER TWO — ASSEMBLY AND RIGGING

21. The angle of attack at which air flowing over the upper surface of an airfoil breaks away and burbles is called the _____ angle of attack.

22. The acute angle that is formed between the chordline of an airplane wing and the airplane's longitudinal axis is called the angle of _____ .

23. An airfoil section that is quite thin is said to have a high _____ ratio.

24. The efficiency of any airfoil is measured in terms of the ratio of the lift produced by the airfoil to the _____ produced.

25. The curvature of the airfoil above and below its chordline is called the _____ of the surface.

26. The ratio of the length (span) of an airfoil to its width (chord) is called the _____ ratio.

27. The point in an aircraft at which all of the aerodynamic forces are considered to be concentrated is called the _____ of the aircraft.

28. The center of gravity of a properly designed and loaded airplane falls slightly _____ (ahead of or behind) the center of pressure.

29. Aerodynamic lift produced by an airplane wing always acts _____ (parallel or perpendicular) to the relative wind.

30. Aerodynamic drag produced by an airplane wing always acts _____ (parallel or perpendicular) to the relative wind.

31. Three types of aerodynamic drag acting on an airfoil are

    A. _____

    B. _____

    C. _____

32. The air that spills over the wing tips of an airplane in flight produces swirling streams of air called wing tip _____ .

AIRFRAME SECTION WORKBOOK

ASSEMBLY AND RIGGING — CHAPTER TWO

33. The axis of an airplane that extends lengthwise through the fuselage from front to rear is called the _____ axis.

34. The axis of an airplane that extends from wing tip to wing tip is called the _____ axis.

35. The axis of an airplane that extends through the center of the fuselage from top to bottom is called the _____ axis.

36. Rotation of an aircraft about its longitudinal axis is called _____ .

37. Rotation of an aircraft about its lateral axis is called _____ .

38. Rotation of an aircraft about its vertical axis is called _____ .

39. The characteristic of an aircraft that causes it to fly hands-off in a straight and level flight path is called _____ .

40. The characteristic of an aircraft that allows it to be directed along a desired flight path is called _____ .

41. The characteristic of an aircraft that causes it to move in response to the pilot's commands is called _____ .

42. An aircraft which tends to return to level flight after it has been disturbed from this flight condition is said to have positive _____ (static or dynamic) stability.

43. If the restorative forces of static stability increase instead of decrease, the aircraft is said to have _____ (positive or negative) dynamic stability.

44. The primary aerodynamic surface that is used to provide longitudinal stability to an airplane is the _____ .

45. An airplane with a long fuselage normally has _____ (more or less) directional stability than an airplane with a short fuselage.

46. Dihedral in an airplane wing is used for _____ stability.

47. When a wing is swept back, the dihedral is effectively _____

CHAPTER TWO — ASSEMBLY AND RIGGING

(increased or decreased).

48. The aileron moving downward travels a _____ (greater or smaller) distance than the aileron moving upward.

49. When spoilers are used for lateral control, the spoiler rises up from the surface of the wing on which the aileron is moving _____ (upward or downward).

50. Fixed spoilers are sometimes installed on the leading edge of a wing near the root. These spoilers cause the center section of the wing to stall out _____ (before or after) the section of the wing that contains the ailerons.

51. When an airplane moves toward the outside of a turn because of too little bank angle, it is said to be _____ (slipping or skidding).

52. A single-piece movable horizontal tail surface of an airplane is called a _____.

53. The movable controls on a V-tail airplane are called _____.

54. A movable tab used to cause an airplane to fly straight and level, hands-off, under a condition of temporary unbalance is called a _____ tab.

55. A servo tab moves in the _____ (same or opposite) direction as the control surface on which it is mounted.

56. A balance tab moves in the _____ (same or opposite) direction as the control surface on which it is mounted.

57. A tab that deflects only when the force needed to move a control surface becomes excessive is called a _____ tab.

58. Two characteristics of an airplane wing increase when Fowler flaps are lowered; these are

   A. _____

   B. _____

ASSEMBLY AND RIGGING

CHAPTER TWO

59. The layer of air that flows immediately over a wing surface is called the _____ layer.

60. Forward flight of a helicopter is attained by tilting the _____ of the rotor forward.

61. The force that causes the fuselage of a helicopter to try to rotate in the direction opposite to that of the main rotor is called _____.

62. The foot-operated pedals in a single-rotor helicopter control the pitch of the _____ rotor.

63. The cause of a force in a helicopter rotor reacting at a point 90 degrees from the point of application, in the direction of rotor rotation, is called gyroscopic _____.

64. The blade of a helicopter rotor whose tip is moving in the same direction the helicopter is moving through the air is called the _____ blade.

65. The blade of a helicopter rotor whose tip is moving in the opposite direction the helicopter is moving through the air is called the _____ blade.

66. When a helicopter is in forward flight, the advancing rotor blade produces more lift than the retreating blade. This difference in lift is called _____ of lift.

67. Helicopter rotor blades are made to flap to help minimize _____ of lift.

68. The tips of some helicopter rotor blades are moved back and forth in their plane of rotation by pivoting about their _____ hinge.

69. A helicopter rotor blade that is free to flap, drag, and feather is called a fully _____ rotor.

70. The coning angle of the blades of a helicopter rotor is caused by a balance between the lift produced by the blade and _____ force acting on the blade.

CHAPTER TWO — ASSEMBLY AND RIGGING

71. The additional lift produced by the downwash of a helicopter rotor system when it is operating in its ground effect extends upward for about _____ (what part) of the rotor disk diameter.

72. A helicopter rotor will continue to turn even if the engine is disconnected from it. The force that causes it to continue to rotate is called an _____ force.

73. The three axes of a helicopter are the

   A. _____

   B. _____

   C. _____

74. The helicopter flight control that changes the pitch of all of the rotor blades at the same time is the _____ pitch control.

75. The helicopter flight control that changes the pitch of the rotor blades as each one reaches a particular point in its rotation is called the _____ pitch control.

76. Engine power of a helicopter is synchronized with the _____ (collective or cyclic) pitch control.

77. The speed of sound varies with the _____ (pressure or temperature) of the air.

78. When an airplane is operating at a speed in which all of the airflow over the surfaces is moving at a speed slower than the speed of sound, it is said to be operating in _____ flight.

79. When an airplane is operating at a speed in which part of the air is moving over the surface slower than the speed of sound while other air is moving at a speed greater than the speed of sound, it is said to be operating in _____ flight.

80. When an airplane is operating at a speed in which all of the airflow over the surfaces is moving at a speed greater than the speed of sound, it is said to be operating in

AIRFRAME SECTION WORKBOOK

ASSEMBLY AND RIGGING

_____ flight.

81. The density of air increases as it flows through a contracting tube when it is moving at a _____ (subsonic or supersonic) rate.

82. When air passes through an oblique shock wave, it slows down to a speed that is _____ (above or below) the speed of sound.

83. When air passes through a normal shock wave, it slows down to a speed that is _____ (above or below) the speed of sound.

84. When air passes through an expansion wave, it _____ (speeds up or slows down).

85. Devices mounted on the surfaces of some airfoils for the purpose of holding high-energy air on the surface and delaying shock-induced separation are called _____ generators.

86. The leading edges of high-speed airplane wings are made sharp to prevent the formation of _____ (normal or oblique) shock waves.

87. Vortex generators mounted on the underside of the horizontal stabilizer improve _____ (high speed or low speed) flight characteristics.

88. A flight control system that allows the pilot to move the surface, but does not allow aerodynamic forces acting on the surfaces to move the cockpit control, is called an _____ control system.

89. An irreversible control system must have some form of artificial _____ so the pilot can sense the amount of load on the control surfaces.

90. One problem of flight at supersonic speeds is aerodynamic heating. The greatest temperature rise is produced at the stagnation points where the air velocity is _____ (greatest or least).

AIRFRAME SECTION WORKBOOK

CHAPTER TWO
ASSEMBLY AND RIGGING

91. Three types of control systems used in aircraft are

    A. _____

    B. _____

    C. _____

92. An aircraft control cable that has indication of internal corrosion should be _____.

93. Tension of aircraft control cables is adjusted with _____.

94. Power-assisted control systems using hydraulic actuators _____ (are or are not) irreversible systems.

95. A fairlead should not be used in a control system to change the direction of cable travel more than _____ (how many) degrees.

96. When control cables pass through the bulkhead of a pressurized compartment, they are routed through _____.

97. A rod-end bearing should be connected to a control arm with the flange of the bearing on the side of the _____ (arm or nut).

98. It is normal practice when rigging control surfaces for the stop on the surface to be reached _____ (before or after) the stop on the cockpit control.

99. Control-cable-tension regulators are installed in the control systems of some large aircraft to compensate for the dimensional changes in the aircraft caused by _____ changes.

100. Using the typical cable-rigging chart on page 71 of the text, find the rigging load (tension) in pounds you would use for a 3/16-inch, 7 x 19 control cable on this particular airplane when the ambient temperature is 90 degrees Fahrenheit. You should use _____ pounds.

101. Control surface movement is checked by using a universal _____ protractor.

AIRFRAME SECTION WORKBOOK

ASSEMBLY AND RIGGING — CHAPTER TWO

102. Wing incidence and dihedral are checked by using a special fixture and a _____ level.

103. When using rigging pins for adjusting an aircraft control system, the turnbuckles should be tightened to get the correct cable tension, and the rigging pin should be _____ (loose or tight) in its hole.

104. Helicopter rotor blade track can be checked with a tracking flag or with a strobe light and _____ installed on the blade tips.

105. An aircraft control surface is most likely to flutter in some flight conditions if it is out of balance with most of the weight _____ (ahead of or behind) the hinge line.

CHAPTER 3

AIRCRAFT FABRIC COVERING

1. The two most important organic fabrics used for covering aircraft structures are

    A. _____

    B. _____

2. Three of the most popular heat-shrinkable fabrics used for covering aircraft structures are

    A. _____

    B. _____

    C. _____

3. The strength of the fabric required for covering an aircraft structure is determined by the aircraft's

    _____ speed and its

    _____ loading.

4. New Grade A cotton fabric has a minimum tensile strength of _____ pounds per inch of width.

5. Grade A cotton fabric has between _____ and _____ threads per inch in both the warp and fill directions.

6. Standard-weight Dacron fabric is _____ (stronger or weaker) than Grade A cotton fabric.

7. Medium-weight Dacron fabric _____ (may or may not) be used as a direct replacement for Grade A cotton fabric.

8. Glass fabric used for covering aircraft structures has a weight of _____ ounces per square yard.

AIRCRAFT FABRIC COVERING — CHAPTER THREE

9. Surface tape is applied over all seams, corners, edges, and places where wear is likely to occur. It is applied at the time the _____ (which one) coat of clear dope is brushed on.

10. Surface tape doped over the trailing edge of wings and control surfaces should be _____ inches wide.

11. A heavy cloth tape put over wing ribs to keep the rib lacing cord from pulling through the covering fabric is called _____ tape.

12. Cotton thread used for hand sewing aircraft fabric must have a tensile strength of at least _____ pounds per single strand.

13. Hand sewing thread and rib lacing cord should both be waxed with _____ (beeswax or paraffin).

14. When self-tapping screws are used to attach fabric to a wing rib, at least _____ (how many) threads should extend through the capstrip.

15. When self-tapping screws are used to attach fabric to a wing rib, a washer, made of _____, should be used under the screw heads.

16. The two types of machine-sewed seams preferred for joining two pieces of aircraft fabric are
    A. _____
    B. _____

17. Machine-sewed seams in aircraft fabric should have between _____ and _____ stitches per inch.

18. When hand sewing aircraft fabric to close the final opening in the covering, a hem should be turned back so the sewing will be done through a double thickness of fabric. The hem should be _____ inch wide.

19. Aircraft fabric should be hand sewed using a baseball stitch, and it should be lock-stitched every _____ inches.

CHAPTER THREE　　　　　　　　　　　　　　　　AIRCRAFT FABRIC COVERING

20. In a hand-sewed seam, there should be at least _____ stitches per inch.

21. A double-stitched lap joint in aircraft fabric should be covered with surface tape that is _____ inches wide.

22. Notches should be cut in the surface tape that is used to cover the trailing edge seams in aircraft fabric if the aircraft has a never-exceed airspeed of more than _____ miles per hour.

23. If a chordwise sewed seam is placed over a wing rib, the rib lacing _____ (should or should not) pass through the seam.

24. A lapped and doped spanwise seam may be made along the leading edge of a wing if the fabric laps by at least _____ inches and the seam is covered with pinked-edge surface tape that is at least _____ inches wide.

25. A lapped and doped spanwise seam may be made along the trailing edge of a wing if the fabric laps by at least _____ inches and the seam is covered with pinked-edge surface tape that is at least _____ inches wide.

26. Wooden structural parts that are likely to contact the doped fabric when it is being applied should be dope-proofed with dope-proof paint, cellulose tape, or _____ foil.

27. All metal edges that are likely to chafe through fabric installed over them should be covered with tape that will not absorb _____; this is called a nonhygroscopic tape.

28. Wing ribs may be braced with a continuous strip of _____ tape wrapped diagonally between the top and bottom rib capstrips of each adjacent rib, midway between the front and rear spars.

29. Before covering a plywood surface with aircraft fabric, the wood must be thoroughly cleaned, primed, and given two brush coats of _____.

AIRCRAFT FABRIC COVERING — CHAPTER THREE

30. Two methods of covering an aircraft structure with fabric are

    A. _____

    B. _____

31. When using the blanket method of covering an aircraft wing, it is preferred that the seams run _____ (chordwise or spanwise).

32. If cotton or linen fabric is to be cemented to an aircraft structure, the structure should be brushed with several coats of full-bodied, clear _____ dope before the fabric is put on.

33. After cotton or linen fabric is attached to a structure, it is shrunk with _____ to remove all of the wrinkles.

34. Dacron fabric is shrunk on the structure with _____ .

35. When shrinking Dacron fabric on an aircraft structure, use _____ (how many) passes with the iron or heat gun.

36. The correct temperature for shrinking Dacron fabric on an aircraft structure is _____ degrees Fahrenheit.

37. A Dacron-covered surface should be doped with a _____ dope.

38. When covering a wing with the envelope method, the machine-sewed seam should be placed along the _____ (leading or trailing) edge.

39. Airplanes with a never-exceed speed in excess of 250 mph should have anti-tear strips placed under the reinforcing tape on the top of the wing ribs, and also on the bottom of the ribs that are in the _____.

40. An airplane with a placarded never-exceed speed of 150 mph should have rib stitches spaced every _____ inches inside the slipstream and every _____ inches outside of the slipstream.

41. For determining rib stitch spacing, the slipstream is considered to be as wide as the diameter of the

CHAPTER THREE AIRCRAFT FABRIC COVERING

propeller plus _____ (how many) extra rib space/spaces on each side.

42. Lengths of waxed rib lacing cord can be joined by using a _____ knot.

43. The spacing between the first and second stitch when rib-stitching a wing should be _____ the normal spacing.

44. The correct knot for tieing off rib stitching is a modified _____ knot.

45. Rib stitch tie-off knots _____ (should or should not) be pulled back through the lacing holes to improve the appearance of the stitching.

46. When re-covering or repairing a control surface, you must be aware that flutter may be caused if weight is added _____ (ahead of or behind) the hinge line.

47. Small cuts and tears in aircraft fabric are hand-sewed using a _____ stitch.

48. When using a baseball stitch, the stitches should be no more than _____ inch apart, and they should extend _____ inch into the untorn material.

49. Baseball stitches should be locked every _____ to _____ stitches and at the ends with a modified seine knot.

50. A doped-on patch over a sewed-up tear should extend at least _____ inches beyond the tear in all directions.

51. When preparing a fabric surface to receive a doped-on patch, the finish should be removed down to the _____ dope.

52. A sewed-in-patch repair can be made to a fabric-covered surface if the damage is no longer than _____ inches in any direction.

53. If a sewed-in-patch repair is laced to a wing rib using new reinforcing tape, the old reinforcing tape and rib stitching _____ (should or should not) be removed.

54. When using a sewed-in-panel repair on a wing, the new fabric should extend _____ inches beyond the ribs adjacent to the damage.

AIRFRAME SECTION WORKBOOK

AIRCRAFT FABRIC COVERING — CHAPTER THREE

55. An unsewed, doped-on patch may be used to repair damage to an aircraft structure if the never-exceed speed is not greater than _____ mph and the damage does not exceed _____ inches in any direction.

56. When making an unsewed, doped-on patch over a hole that measures up to eight inches in any direction, the fabric should extend beyond the damage for one fourth of the hole dimension or a maximum of _____ inches.

57. When using a doped-in-panel repair for the top of the wing, the fabric should extend from around the trailing edge, over the leading edge, back to the _____.

58. A doped-in-panel repair should extend _____ inches beyond each rib adjacent to the damaged area.

59. An alternate attachment of a doped-in-panel repair over a metal leading edge is to lap the new panel at least _____ inches over the old fabric at the nose of the leading edge and finish it with pinked-edge surface tape at least _____ inches wide.

60. The new panel of a doped-in-panel repair should be given its first coat of dope _____ (before or after) the edges of the panel have been laced to the ribs.

61. Fungus inhibitor is mixed with dope used on a cotton-covered aircraft structure and applied with the _____ (which one) coat.

62. The dope coat containing the fungus inhibitor should be applied by _____ (brushing or spraying).

63. Ultraviolet light is kept from damaging the clear dope and the fabric by using two coats of dope mixed with _____ paste or powder.

64. When making a fabric test on an aircraft structure, the test should be made in a section of the fabric that is finished with a _____ (light or dark) color.

CHAPTER THREE — AIRCRAFT FABRIC COVERING

65. A fabric punch test _____ (is or is not) as accurate as a pull test for determining the strength of aircraft fabric.

66. For aircraft fabric to be considered airworthy, it must have at least _____ percent of the strength of the fabric that was originally installed on the aircraft by the manufacturer.

67. When Grade A cotton fabric installed on an aircraft with a never-exceed speed of 150 mph and a wing loading of 6.5 psf deteriorates to a strength of 50 pounds per inch, it is considered _____ (airworthy or unairworthy).

68. Surface tape, patches, inspection rings, and drainage grommets are installed as you apply the _____ (which one) coat of clear dope.

69. It is extremely important when dry-sanding a doped aircraft surface that the component be electrically grounded to prevent sparks caused by _____ electricity.

70. Conditions of high humidity during doping will usually cause the dope to _____.

71. Two types of aircraft dope are

   A. _____

   B. _____

72. Of the two types of dope in use _____ (which one) is the more flammable.

73. When mixing aluminum paste with clear dope, you should mix about _____ pounds of aluminum paste with five gallons of clear dope.

CHAPTER 4

AIRCRAFT PAINTING AND FINISHING

1. Acetone should not be used as a dope thinner because it evaporates too _____ (quickly or slowly).

2. Shellac can be thinned with _____ alcohol.

3. One use of toluene is as a thinner for _____ (what kind) primer.

4. The six main constituents of aircraft dope are

   A. _____
   B. _____
   C. _____
   D. _____
   E. _____
   F. _____

5. Nitrocellulose lacquer _____ (can or cannot) be applied over zinc chromate primer.

6. A wash primer is a _____ (one or two)-part primer.

7. Modified zinc chromate primer _____ (is or is not) satisfactory for application over bare metal.

8. Interior wood parts of an aircraft are usually finished with several coats of clear _____.

9. Enamel is removed from a surface by the use of a paint stripper that causes the enamel film to _____ and pull away from the surface.

AIRCRAFT PAINTING AND FINISHING — CHAPTER FOUR

10. Paint should be removed from a fiberglass component with _____ (paint stripper or sandpaper).

11. The finish that is used on most modern aircraft is a _____ enamel.

12. When mixing an epoxy finishing material, you should add the _____ (resin or converter) to the _____ (resin or converter).

13. After mixing the converter and resin, the mixture should be allowed to stand for about _____ minutes before it is used.

14. Polyurethane enamel has a _____ (long or short) flowout time.

15. For best results, polyurethane enamel should be sprayed within _____ hours after it is mixed.

16. When thinning a finishing material for spraying, you can get consistent results if the viscosity of the mixed material is checked with a viscosity _____ .

17. When removing masking tape from a freshly painted surface, it should be pulled _____ (parallel or perpendicular) to the surface.

18. Three types of decals used for use on aircraft are

    A. _____

    B. _____

    C. _____

CHAPTER 5

AIRCRAFT STRUCTURAL REPAIRS

1. The final authority for aircraft structural repairs is the aircraft _____ .

2. Any repair to an aircraft structure must restore all of the _____ lost when the structure was damaged.

3. The minimum corner radius used in a rectangular cutout in an aircraft structure is _____ inch.

4. The replacement of damaged sheet metal with a thinner gage of stronger material _____ (is or is not) normally considered to be a good practice.

5. When choosing a material for a structural repair, the tensile strength and the _____ strength of the material must both be considered.

6. If no other information is available, rivet spacing used for a wing repair can be the same as that used in the nearest riveted seam _____ (inboard or outboard) of the damaged area.

7. The typical choice for a rivet diameter used to join two pieces of 0.051 aluminum alloy sheet is _____ inch.

8. Using the charts on pages 128 and 129 of the text, find the number of 1/8-inch 2117T rivets you would need to restore the strength lost by a two-inch-long crack in a piece of 0.040 aluminum alloy sheet. You would need _____ (how many) rivets on each side of the crack.

9. A small damage to an aircraft part that does not cause any flight restrictions and that can be corrected by simple procedures is called _____ damage.

AIRCRAFT STRUCTURAL REPAIRS — CHAPTER FIVE

10. The five basic stresses that are encountered in aircraft structure are

    A. _____

    B. _____

    C. _____

    D. _____

    E. _____

11. The primary stress in the upper wing skin of an airplane in flight is _____.

12. The primary stress in a wing strut in flight is _____.

13. The stress that tries to cause one layer of a material to slide over another layer is a _____ stress.

14. A torsional stress is made up of two other stresses acting at right angles to each other; these stresses are

    A. _____

    B. _____

15. The most generally used devices for temporarily holding sheet metal parts together while an aircraft is being repaired are _____ fasteners.

16. Throatless shears _____ (can or cannot) be used to cut across a wide sheet of metal.

17. The maximum thickness of metal that should be cut with a Ketts saw is _____ inch.

18. The most widely used type of drill motor for aircraft repair is _____ (electrically or pneumatically) operated.

19. The most widely used grinding wheel for cutting steel is made of _____.

20. A bar folder is used for making bends in the edges of _____

CHAPTER FIVE　　　　AIRCRAFT STRUCTURAL REPAIRS

(thick or thin) sheet metal.

21. The proper tool for making a right-angle bend across the middle of a piece of metal is a _____ brake.

22. The shop tool used for making a large-radius bend in a piece of sheet metal is a _____ former.

23. Large sheets of metal are formed into compound curves in an aircraft factory by the use of _____ fitted with male and female dies of the correct shape.

24. If, when forming a piece of aluminum alloy, a bend is made incorrectly, it _____ (is or is not) permissible to straighten the material and remake the bend.

25. The correct device for marking aluminum alloy sheets when making a sheet metal layout is a _____ .

26. The actual amount of material used in a bend in sheet metal is called the _____ .

27. Four things that determine the bend allowance for a piece of sheet metal are

　　A. _____
　　B. _____
　　C. _____
　　D. _____

28. When bending a piece of sheet metal, the material on the inside of the bend shrinks and that on the outside of the bend stretches. Material near the center keeps its original dimensions. The line that does not change is called the _____ line of the material.

29. Bend radius used in problems of sheet metal layout is always the radius of the _____ (inside or outside) of the bend.

AIRFRAME SECTION WORKBOOK　　　　ASA/FIVE-3

AIRCRAFT STRUCTURAL REPAIRS

30. The bend allowance for a 90-degree bend in a piece of 0.051-inch sheet aluminum alloy using a 1/4-inch bend radius is _____ inch.

31. Using the bend allowance chart on page 149 of the text, find the amount of material needed for a 90-degree bend in a piece of 0.040-inch aluminum alloy using a 1/4-inch bend radius. Bend allowance is _____ inch.

32. The line formed by the intersection of lines projected along the outside of a curved sheet metal part is called the _____ line.

33. The distance between the bend tangent line and the mold line of a part being formed is called _____ .

34. For a 90-degree bend in a piece of sheet metal, setback is equal to the metal thickness plus the _____ .

35. Using the K-chart on page 150 of the text, find the amount of setback needed for a 135-degree bend in a piece of 0.040-inch aluminum alloy using a bend radius of 1/4 inch. Setback for this bend is _____ inch.

36. A line vertically upward from the edge of the radius bar on the jaw of a cornice brake is called the _____ line.

37. An angle formed when a piece of sheet metal is bent through more than 90 degrees is called a _____ angle.

38. An angle formed when a piece of sheet metal is bent through less than 90 degrees is called an _____ angle.

39. According to the chart in Fig. 5-28, on page 151 of the text, the minimum bend radius for a piece of 0.071-inch 7075-T6 aluminum alloy is _____ inch.

40. The distance between the sight line and the bend tangent line when using a 1/2-inch bend radius is _____ inch.

CHAPTER FIVE
AIRCRAFT STRUCTURAL REPAIRS

41. Find the width of material needed to form a 2-inch-wide channel with each side 2 inches high. The material is 0.051 inch thick and the bend radius is 1/8 inch. The total width of material needed is _____ inches.

42. A hole drilled in sheet metal at the point two bends intersect is called a _____ hole.

43. A commonly used size for a relief hole in a piece of sheet metal is for its diameter to be the same as the _____ .

44. The center of a relief hole is the intersection of the _____ bend tangent lines.

45. Lightening holes stiffen a piece of thin sheet metal when the edges of the hole are _____ .

46. To form a concave curve in a piece of sheet metal by shrinking, you should start working at the _____ (center or ends) of the bend.

47. To form a convex curve in a piece of sheet metal by stretching, you should start working at the _____ (center or ends) of the bend.

48. When forming a streamlined cover plate by bumping it into a female die, you should start hammering the metal _____ (in the center or around the edges) of the die.

49. A small offset formed in a piece of angle strip so it will fit over another member is called a _____ .

50. The drill used for drilling stainless steel should be ground to a _____ (flatter or steeper) angle than is used for drilling mild steel.

51. When drilling stainless steel, the drill should be turned _____ (faster or slower) than it is turned for drilling mild steel.

AIRCRAFT STRUCTURAL REPAIRS — CHAPTER FIVE

52. Sheet magnesium work-hardens _____ (more or less) readily than sheet aluminum.

53. The recommended type of fire extinguisher to use on a magnesium metal fire is a _____ extinguisher.

54. A general rule for selecting the diameter of a rivet to use in a sheet metal structure is for the rivet diameter to be three times the thickness of the _____ (thicker or thinner) sheet being joined.

55. The shank of a rivet should stick through the material being joined by _____ times the diameter of the rivet shank.

56. The shop head on a rivet should have a diameter of _____ times the rivet shank diameter.

57. Edge distance with reference to rivet installation is the distance between the edge of the sheet and the _____ (edge or center) of the rivet hole.

58. The minimum edge distance for a sheet metal splice is _____ rivet diameters.

59. Of these rivets, 2017-T, 2117-T, and 2024-T, only one can be driven as it is received from the supplier. This is the _____ rivet.

60. The distance between the centers of adjacent rivets in the same row is called rivet _____.

61. Rivets in a row should be placed no closer together than _____ shank diameters.

62. The perpendicular distance between rows of rivets is called _____.

63. Transverse pitch of rows of rivets is normally _____ percent of the rivet pitch.

64. The tool used to cut a taper in the edges of a hole for the installation of a flush rivet is called a _____.

65. The process of forming the edges of a hole in thin sheet metal so a flush rivet can be used is called

CHAPTER FIVE — AIRCRAFT STRUCTURAL REPAIRS

_____.

66. The most uniform riveting along the edge of a metal sheet is done with a _____ riveter.

67. The correct size drill to use for installing a 1/8-inch rivet is a number _____ drill.

68. When flush riveting two sheets of metal together, if the thickness of the top sheet is thinner than the head of a flush rivet, the top sheet should be _____ (countersunk or dimpled).

69. Two types of dimpling are

   A. _____

   B. _____

70. Magnesium alloys, titanium alloys, and 7075-T6 aluminum alloy should be _____ (hot or cold) dimpled.

71. When blind riveting on a large aircraft structure, the bucker signals that a rivet should be marked and drilled out by the use of _____ (how many) taps.

72. A bucking bar for use with 1/8-inch rivets should weigh between _____ and _____ pounds.

73. The thickness of a properly formed shop head on a rivet should be _____ the shank diameter.

74. A precision cutting tool used to shave the head of a rivet that has been driven into a countersunk hole to form a flush head is called a _____.

75. A joggled rivet removed from a structure is an indication that the rivet has failed in _____.

76. When a riveted joint fails by the rivets tearing the skin, the joint has failed in _____.

AIRCRAFT STRUCTURAL REPAIRS — CHAPTER FIVE

77. When removing a solid 1/8-inch rivet from an aircraft structure, you should drill through the head with a number _____ drill.

78. Mechanical-lock self-plugging rivets _____ (can or cannot) normally be used to replace solid rivets, size for size.

79. Pull-thru rivets _____ (can or cannot) be used in primary structural repair of an aircraft.

80. A special blind rivet that is used to form a threaded insert in a hole in thin sheet metal is called a _____ .

81. A Rivnut with a small projection on its head to keep it from turning in its hole is called a _____ Rivnut.

82. The correct flat-head Rivnut to install in a piece of 0.064 aluminum alloy for use with a number 10 screw is a 10-_____ .

83. Deutsch rivets are installed in _____ (lightweight or heavyweight) aircraft structure.

84. A Hi-Shear rivet is held in its hole with a metal _____ swaged into a groove in the end of the rivet pin.

85. The shape of patch that gives the best concentration of rivets within the critical stress area is an elongated _____ patch.

86. A common shape for a flush patch in an aircraft skin is a _____ patch.

87. To determine the amount of strength lost when a bulb angle stringer is damaged, you should consider the area of the bulb to be _____ times the thickness of the material in the standing leg.

88. The minimum rivet spacing for making a bulb angle splice is _____ rivet diameters.

89. One type of damage that is likely to be found in the trailing edges of wings and tail surfaces is

CHAPTER FIVE — AIRCRAFT STRUCTURAL REPAIRS

_____ .

90. If infrared lamps are used to speed up the cure of a sealant on an aircraft structure, they should be placed no closer than _____ inches from the sealant.

91. The type of damage honeycomb structure is likely to receive from flight loads or sonic vibration is _____ .

92. The simplest check for delamination in a honeycomb structure is the metallic _____ test.

93. One of the most efficient tools for removing damaged honeycomb core from a laminated panel is a high-speed _____ .

94. A solvent specially suited for cleaning a bonded honeycomb structure to prepare it for repair is _____ .

95. Damaged aluminum honeycomb core material may be replaced with a plug of _____ honeycomb.

96. Damaged bonded honeycomb structure can be repaired by removing the damaged core and filling the hole with a potting compound if the cleaned-out damage does not make a hole with a diameter of greater than _____ inch.

97. A filler material made of millions of tiny bubbles of phenolic resin may be mixed with potting compound to add body with little weight to a repair core. This material is called _____ .

98. Two types of plastic resins are used in aircraft construction; these are

   A. _____

   B. _____

99. A plastic resin material that can be softened by heat, and when the heat is removed, will return to its normal state is called a _____ resin.

AIRCRAFT STRUCTURAL REPAIRS　　　　　　　　　　　CHAPTER FIVE

100. A plastic resin material that is cured by heat, but will not soften when it is later heated is called a _____ resin.

101. The transparent plastic material that is most readily softened by acetone is an _____ (acetate or acrylic) plastic.

102. Masking paper that has hardened on transparent plastic material can be removed by carefully rubbing it with _____ (aliphatic or aromatic) naphtha.

103. Tiny cracks that form in the surface of a transparent plastic material because of heat or other stress are called _____ .

104. The drill used for drilling acrylic plastic material should have a tip angle that is _____ (flatter or steeper) than that used for drilling metal.

105. The most effective method of joining pieces of acrylic plastic is the _____ method.

106. For the maximum strength of a cemented joint in acrylic plastic, the material should be heated to a temperature _____ (above or below) that needed to soften it, and held at this temperature until it is heated evenly throughout.

107. Transparent plastic windshields can be cleaned with mild _____ and plenty of clean water.

108. A material that is satisfactory for masking an acrylic window when an aircraft is being stripped for painting is _____ sheet.

109. In addition to retaining all of the aerodynamic and strength requirements when repairing a radome, you must be careful that its _____ characteristics are not changed.

110. The reference wood for aircraft structure is _____ .

111. White pine _____ (can or cannot) be used as a direct replacement for spruce in aircraft structure.

CHAPTER FIVE　　　　　　　　　　　　　　AIRCRAFT STRUCTURAL REPAIRS

112. The most widely used glue for aircraft structural work is _____ glue.

113. Cabinet maker clamps are used for applying pressure to a glue joint when repairing an aircraft structure. The correct pressure for soft wood is between _____ and _____ psi.

114. The correct slope for a scarf joint used to repair a wood wing spar is _____ to 1.

115. A scarf joint used for splicing a wood wing spar _____ (should or should not) be sandpapered to get a smooth surface for the glue.

116. The best repair for a damaged plywood skin on an aircraft is a _____ (splayed or scarf) patch.

117. The edges of a splayed patch for a plywood skin are tapered to a slope of _____ to 1.

118. The edges of a scarf patch for a plywood skin are tapered to a slope of _____ to 1.

119. A plywood skin on an aircraft can be repaired by covering it with a fabric patch if the cleaned-out hole does not have a diameter of greater than _____ inch.

120. A splice in a wood wing spar _____ (may or may not) extend under a lift-strut fitting.

CHAPTER 6

AIRCRAFT WELDING

1. Three types of welding used for aircraft construction and repair are

    A. _____

    B. _____

    C. _____

2. Two fuel gases used for welding are

    A. _____

    B. _____

3. Aluminum is best gas welded with an _____

    (oxyacetylene or oxyhydrogen) flame.

4. The two most widely used types of electric arc welding for aircraft construction and repair are

    A. _____

    B. _____

5. Three types of electric resistance welding are

    A. _____

    B. _____

    C. _____

6. Acetylene gas becomes unstable when it is stored under pressure greater than

    _____ psi.

7. In order to store acetylene safely under pressure of up to 250 psi, it is dissolved in

    _____.

8. Welding oxygen is stored in steel cylinders under a pressure of _____ psi.

AIRCRAFT WELDING — CHAPTER SIX

9. A suitable lubricant for oxygen equipment and fittings is _____ .

10. Two types of oxyacetylene welding torches are

    A. _____

    B. _____

11. The type of welding torch that has two control valves is the _____ torch.

12. The size of orifice in a welding torch tip determines the _____ (heat or temperature) applied to the work.

13. The temperature of an oxyacetylene flame ranges from approximately _____ to _____ degrees Fahrenheit.

14. An oxyacetylene flame with a rounded inner cone and no feather around it is a _____ flame.

15. An oxyacetylene flame with a definite feather around the inner cone is a _____ flame.

16. An oxyacetylene flame with a short, pointed inner cone is an _____ flame.

17. A momentary backward flow of gases in a torch that causes the flame to go out is called a _____ .

18. The condition in which gases burn inside the torch is called _____ .

19. When shutting down an oxyacetylene welding torch, the _____ (which gas) should be shut off first.

20. For welding heavy materials, a welding torch should be held so the tip of the inner cone is about _____ inch from the metal.

CHAPTER SIX — AIRCRAFT WELDING

21. The torch is pointed in the direction the weld is being made in _____ (forehand or backhand) welding.

22. Before repairing a crack by welding, the ends of the crack should be _____ .

23. Cadmium plating must be removed from a piece of steel before it is welded. This can be done by soaking it in a solution of _____ and water.

24. Chrome molybdenum steel should be welded with a soft, _____ (what type) flame.

25. Gas welding _____ (is or is not) considered to be a satisfactory method for joining stainless steel structural parts.

26. Stainless steel should be welded with slightly _____ (what type) flame.

27. When gas-welding aluminum, it _____ (is or is not) necessary to use flux.

28. Welding flux is removed from a weld in a piece of aluminum by scrubbing it with a solution of _____ acid and water.

29. Aluminum _____ (does or does not) change its color before it melts.

30. Aluminum should be welded with a _____ (what type) flame.

31. Gas welding _____ (is or is not) a good method of joining titanium sheets.

32. When cutting steel with a torch, the metal is preheated with an oxyacetylene flame until it is red hot and then a stream of high-pressure _____ is directed onto the metal.

33. In the process of brazing, the base metal _____ (is or is not) heated to its melting point.

AIRCRAFT WELDING — CHAPTER SIX

34. The molten metal is actually carried into a joint being formed by brazing by _____ action.

35. End fittings are attached to stainless steel lines used to carry oxygen by _____.

36. Soft solder is an alloy of _____ and _____ .

37. The tip of a soldering iron should be cleaned and coated with _____ so it will conduct heat into the metal being soldered.

38. The temperature of an electric arc is approximately _____ degrees Fahrenheit.

39. Metallic arc welding uses a _____ (consumable or nonconsumable) electrode.

40. When welding with a gas-shielded arc, flux _____ (is or is not) required.

41. TIG welding uses a _____ (consumable or nonconsumable) electrode.

42. Two inert gases that can be used when TIG welding are

    A. _____

    B. _____

43. MIG welding uses a _____ (consumable or nonconsumable) electrode.

44. An extremely high-temperature extension of TIG welding is _____ arc welding.

45. Heat-treated members of an aircraft structure _____ (should or should not) be repaired in the field by welding.

46. When partial replacement of a steel tube structure must be made, the _____ (inner or outer) sleeve method is normally preferred.

47. Engine-mount tubing should be repaired by using an _____

(inner or outer) tube repair.

48. Dents in a structural tube member can be repaired by welding patches over them if the dent is not deeper than _____ of the tube diameter, if it does not involve more than _____ of the tube circumference, or if is not longer than _____ tube diameter.

49. When making an inner-sleeve reinforced splice, the ends of the original tube and the replacement tube should be separated by a gap of _____ inch so the two tubes can be welded to the inner tube.

CHAPTER 7

ICE AND RAIN PROTECTION

1. Two types of ice that form on an aircraft structure in flight are

   A. _____

   B. _____

2. Two requirements for ice to form on an aircraft structure in flight are

   A. _____

   B. _____

3. Carburetor ice _____ (can or cannot) form when there is no visible moisture in the air.

4. An ice control system that prevents the formation of ice on an aircraft is called an _____ system.

5. An ice control system that removes ice after it forms is called a _____ system.

6. Pneumatic deicer boots installed on a reciprocating-engine-powered airplane are inflated with air pressure produced by the exhaust of an engine-driven _____ pump.

7. Some deicer boots are attached to wing and empennage leading edges by screws and _____.

8. Most modern deicer boots are attached to the leading edge of the surfaces with _____.

9. When a pneumatic deicer system is not in operation, the boots are connected to the inlet side of the _____ pump to hold them deflated.

10. Most engine-driven vacuum pumps are of the _____ type.

ICE AND RAIN PROTECTION — CHAPTER SEVEN

11. An engine-driven vacuum pump that is lubricated by engine oil passing through it is called a _____ vacuum pump.

12. The vanes of a dry vacuum pump are made of _____ and they do not require any lubrication.

13. Oil in the air discharged from a wet vacuum pump is removed by an _____ and returned to the engine crankcase.

14. The sequence in which the tubes in the deicer boots are inflated and deflated is controlled by a _____.

15. Rubber deicer boots should be cleaned with _____.

16. Thermal anti-icing systems installed on reciprocating-engine-powered airplanes get their heat from _____.

17. Anti-icing systems using exhaust heaters get their heat from _____ tubes which surround the exhaust pipes.

18. Hot air for wing and tail anti-icing on turbine-engine-powered airplanes comes from the engine _____.

19. Expansion and contraction of the ducts in a thermal anti-icing system are allowed for by the use of stainless steel _____.

20. Deicing fluids sprayed on an aircraft before takeoff are normally composed of _____ and _____.

21. If ice has formed inside a turbine engine so the compressor cannot be turned by hand, it should be freed with _____ before any attempt is made to start the engine.

22. Ice is kept from forming on the windshield of modern jet airplanes by _____ coatings between the plies of glass.

CHAPTER SEVEN
ICE AND RAIN PROTECTION

23. The electricity used to produce the heat inside the windshield of a modern jet airplane is _____ (AC or DC).

24. Small surface scratches _____ (should or should not) be polished out of a tempered-glass windshield.

25. The windshields and carburetors on some aircraft are deiced by a spray of _____ .

26. Ice is kept from forming on the pitot tube by an _____ heater installed in the tube.

27. Three methods of keeping water from obstructing the pilot's visibility during flight through heavy rain are

   A. _____
   B. _____
   C. _____

28. The chemical rain repellant system _____ (should or should not) be operated on a dry windshield.

CHAPTER 8

HYDRAULIC AND PNEUMATIC POWER SYSTEMS

1. Three types of fluid that may be used in aircraft hydraulic systems are

    A. _____

    B. _____

    C. _____

2. Natural rubber seals are used in hydraulic systems using _____-base fluid.

3. Vegetable-base hydraulic fluid is made of _____ oil and _____ .

4. Mineral-base hydraulic fluid is dyed _____ .

5. The principal advantage of phosphate ester-base hydraulic fluids is that they are _____ .

6. Vinyl insulation on electrical wiring may be damaged if hydraulic fluid with a _____ base is spilled on it.

7. The element in a micronic hydraulic filter is made of _____ and it is discarded rather than cleaned.

8. Filter effectiveness is measured in microns. One micron is _____ of an inch.

9. Hydraulic filters have a _____ valve in them that allows unfiltered fluid to pass through the system if the filter should become clogged.

10. The four basic components in a simple hydraulic system are

    A. _____

    B. _____

HYDRAULIC AND PNEUMATIC POWER SYSTEMS — CHAPTER EIGHT

C. _____

D. _____

11. The device in a hydraulic system that allows fluid to be stored under pressure is the

    _____ .

12. A device in a hydraulic system that allows fluid to flow in one direction, but prevents its flowing in

    the opposite direction, is a _____ valve.

13. Some aircraft fly at such high altitudes that there is not enough air pressure to force hydraulic fluid

    from the reservoir to the inlet of the pump. The reservoirs used on these aircraft are

    _____ .

14. Three methods of pressurizing a hydraulic reservoir are

    A. _____

    B. _____

    C. _____

15. Hydraulic reservoirs that supply fluid to both normal and emergency systems have a standpipe

    around the exit that feeds the _____

    (normal or emergency) system.

16. A hydraulic hand pump that moves fluid when the handle is moved in both directions is called a

    _____ pump.

17. Two basic types of power-driven hydraulic pumps are

    A. _____

    B. _____

18. Fluid moving through a gear-type pump passes _____

    (between or around) the gears.

19. The gerotor-type pump shown in Fig. 8-11 on page 319 of the text is essentially a four-tooth

    external spur gear turning inside a _____-tooth internal gear.

CHAPTER EIGHT
HYDRAULIC AND PNEUMATIC POWER SYSTEMS

20. Vane-type pumps are used to move fluid when the system requires _____ (high or low) volume and _____ (high or low) pressure.

21. Piston-type pumps are used when a _____ (high or low) pressure is needed.

22. A jammed hydraulic pump is kept from damaging the engine by a _____ section in the pump drive coupling.

23. The angular-type piston pump shown in Fig. 8-16 on page 321of the text is a _____ (constant or variable)-delivery pump.

24. If a pressure relief valve were used as a pressure regulator, it _____ (would or would not) keep a constant load on the pump.

25. Another name for a system pressure regulator is an _____ valve because it keeps the load off of the engine-driven pump when no units in the system are demanding fluid.

26. Three types of accumulators that may be used in a hydraulic system are

    A. _____

    B. _____

    C. _____

27. Accumulators are preloaded with compressed air or _____ .

28. A check valve that allows full flow of fluid in one direction, but restricts the flow in the opposite direction, is called an _____ check valve.

29. Line disconnect valves are installed in the flexible fluid lines that connect the hydraulic _____ (which component) into the system.

30. An actuating cylinder that uses hydraulic fluid to move the piston in one direction, but uses a spring to return the piston, is called a _____ actuating cylinder.

31. An actuating cylinder that uses a piston with the same area on each side is called a _____ actuator.

HYDRAULIC AND PNEUMATIC POWER SYSTEMS — CHAPTER EIGHT

32. Two types of closed-center selector valves used in aircraft hydraulic systems are

    A. _____

    B. _____

33. Some aircraft hydraulic systems have back-up systems for emergency extension of the landing gear and the application of brakes that are actuated by _____ .

34. High-pressure pneumatic systems have two-stage or three-stage compressors that are engine driven or driven by an _____ or _____ motor.

35. Medium-pressure pneumatic systems installed on turbine-engine-powered aircraft get their air from the engine _____ .

36. Low-pressure pneumatic systems used on reciprocating-engine aircraft get their air from _____ - type engine-driven pumps.

37. The speed of actuation of a pneumatic system component is normally controlled by a variable _____ .

38. The main hydraulic system and the emergency system for the application of brakes or for lowering the landing gear meet at _____ valves located on the component to be actuated.

39. The high-pressure air compressor shown in the pneumatic system in Fig. 8-43 on page 338 of the text is driven by a _____ motor.

40. Moisture that condenses from the air is kept from freezing by an electric _____ blanket around the moisture separator.

CHAPTER 9

LANDING GEAR SYSTEMS

1. A landing gear with more than two wheels attached to one strut is called a _____ landing gear.

2. An aircraft shock absorber that uses oil and air to absorb landing and taxi shocks is generally called an _____ strut.

3. The rate of oil flow between the lower and upper cylinders in an oleo strut is controlled by a tapered _____ pin.

4. The rebound of some oleo struts is minimized by the use of a _____ device inside the strut.

5. Retractable nose gear shock struts have a built-in _____ cam to straighten the wheel before it folds into the wheel well.

6. When servicing an oleo strut with oil and air, the strut is completely deflated, the air valve is removed, and the strut filled with oil to the level of the _____ opening.

7. After an oleo strut is filled with oil, the air valve is replaced and the strut is charged with compressed air or _____ .

8. The proper amount of air in an oleo strut on most large aircraft is determined by measuring the amount the _____ extends when the weight of the aircraft is on the landing gear.

9. Air trapped in the oil chamber of an oleo strut can be removed by _____ the strut. This is done by moving the strut up and down while a hose connects the strut to a container of fluid.

LANDING GEAR SYSTEMS — CHAPTER NINE

10. The devices on an oleo strut that keep the wheel aligned with the aircraft are called _____.

11. The device on a retractable landing gear that attaches the gear to the aircraft structure and allows it to pivot when retracting is called a _____.

12. The component in a landing gear that keeps the wheel from moving backward and forward is called a _____ strut.

13. Many retractable landing gears are folded into the aircraft structure by the actuator causing the _____ strut to fold in its center.

14. A hydraulically actuated retractable landing gear has devices that keep the wheel-well doors from closing until the landing gear is fully retracted. These devices are called _____ valves.

15. A safety switch keeps the landing gear from being accidentally retracted when the aircraft is on the ground. This switch is usually actuated by the _____ on the main landing gear.

16. One type of ground lock for an aircraft landing gear is a pin that fits through holes in the drag strut or side strut that are aligned when the landing gear is locked down. To keep this pin from accidentally being left in when the aircraft is prepared for flight, a _____ is attached to it.

17. Most modern aircraft use _____ (what color) indicator lights to show that the landing gear is down and locked.

18. Nose wheel steering is normally done on large aircraft by hydraulic actuators called _____ cylinders.

19. Vibration of the nose wheel is prevented by a _____ between the shock strut cylinder and piston.

CHAPTER NINE
LANDING GEAR SYSTEMS

20. A shimmy damper allows the nose wheel to move slowly for steering, but it acts as a snubber when the movement is _____ .

21. Some shimmy dampers serve the dual purpose of reducing shimmy and of providing nose wheel steering. These units are called _____ dampers.

22. Three types of brake systems used on aircraft are

    A. _____

    B. _____

    C. _____

23. When the brake is off, the wheel cylinder in an independent brake system is vented to the brake reservoir through the _____ port in the brake master cylinder.

24. Hydraulic pressure for emergency actuation of hydraulic power brakes is stored in a brake system _____ .

25. A power brake control valve is essentially a pressure _____. The harder the pilot pushes on the brake pedal, the more pressure is directed to the brakes.

26. The volume of fluid sent into the brake wheel cylinders in a power brake system is increased by the use of brake _____ cylinders.

27. A brake debooster cylinder _____ (increases or decreases) the pressure sent to the brake wheel cylinders.

28. The force applied by the pilot on the brake pedals is increased by hydraulic system pressure in a _____ brake system.

29. On airplanes that have nose wheel braking, the nose wheel brake will not actuate when the airplane is moving below about _____ mph.

30. The brake assemblies installed on most small airplanes are of the _____ _____ type.

LANDING GEAR SYSTEMS — CHAPTER NINE

31. When more braking action is needed than is available with a single disk brake, a _____ brake may be used.

32. Most heavy aircraft using power brake systems are equipped with _____ brakes.

33. The high-friction brake lining used in a segmented rotor brake is mounted on the _____ (stator or rotor) assembly.

34. An expander tube brake is applied when hydraulic pressure expands a fabric-reinforced _____ (what material) tube and forces the brake blocks out tightly against the brake drum.

35. Two methods of bleeding aircraft hydraulic brakes are

   A. _____

   B. _____

36. You know that hydraulic brakes need bleeding when the pedal has a _____ feel.

37. The most popular type of aircraft wheel is the _____ wheel.

38. Air is kept from leaking between the halves of a split wheel when a tubeless tire is mounted on it by the use of a _____ between the wheel halves.

39. Excessive application of the brakes can overheat a wheel so much that the tire is likely to explode. In the event of such overheating, _____ relief plugs in the wheel will melt and deflate the tire safely before the pressure builds up high enough to cause it to explode.

40. The bearings used in aircraft wheels are of the _____ type.

41. Aircraft tires are designed to flex _____ (more or less) than automobile tires.

42. Excessive wear on the center of the tire tread is an indication that the tire has operated in an _____ (overinflated or underinflated) condition.

43. Excessive wear on the edges of the tire tread is an indication that the tire has operated in an _____ (overinflated or underinflated) condition.

44. Aircraft tires should not be checked for inflation pressure until they have cooled at least _____ (how many) hours after a flight.

45. A newly-mounted nylon-reinforced tire will stretch and drop the inflation pressure by as much as _____ to _____ percent in the first 24 hours after mounting.

46. The air pressure in dual-mounted tires must be kept the same, and any difference of more than _____ psi should be noted in the aircraft logs so these tires can be carefully watched.

47. If an airplane is to remain out of service for a long period of time, it should be moved about every _____ hours to rotate the wheels, or the weight should be taken off of the tires.

48. It _____ (is or is not) a good practice to recap airplane tires.

49. Weather cracks in the sidewall rubber of a tire that extends into the cords _____ (is or is not) a reason for rejecting the tire.

50. A tire mounted on a wheel that has been overheated enough to melt the thermal fuse _____ (should or should not) be scrapped.

51. When mounting a tubeless tire on an airplane wheel, you may lubricate the beads of the tires with a _____ and water solution.

52. Balance marks on an aircraft tube identify the _____ (heavy or light) point of the tube.

53. Balance marks on an aircraft tire identify the _____ (heavy or light) point of the tire.

LANDING GEAR SYSTEMS — CHAPTER NINE

54. After the initial inflation of a newly-mounted tube-type tire, the tire should be completely _____ and then reinflated with the correct pressure.

55. If an aircraft tire is going to be removed from the wheel, the tire should be deflated _____ (before or after) it is removed from the aircraft.

56. The halves of split wheels are each marked with the letter "L." When assembling these wheels the marks should be _____ (next to or across the wheel from) each other.

57. The best type of pressure gage to use for maintaining aircraft tire pressure is a _____-type gage.

58. Aircraft tubes should be stored in their _____.

59. The skid control generator for an antiskid system is mounted in the wheel _____.

60. Hydraulic pressure to the brake of a wheel that is slowing down fast enough to indicate that a skid is likely is released by the _____ valve.

61. An antiskid system does not operate when the weight of the aircraft is on the wheels and the aircraft is moving at a speed of less than _____ to _____ mph.

62. A retraction test of the landing gear with the aircraft on jacks _____ (is or is not) part of an annual inspection.

CHAPTER 10

FIRE PROTECTION SYSTEMS

1. An area or region in an aircraft that the manufacturer determines should have a fire-detection and/or fire-extinguishing system is called a fire _____ .

2. The three types of fire-detection systems most commonly used in aircraft are

   A. _____

   B. _____

   C. _____

3. A thermal switch fire-detection system _____ (will or will not) signal a general overheat condition.

4. A thermal switch fire detection system _____ (will or will not) signal the presence of a fire even though the test switch shows a break in the wire connecting the switches together.

5. A break in either of the loops of wiring used in a Fenwal spot-detector circuit _____ (will or will not) cause the system to fail to detect a fire.

6. A thermocouple fire-detection system is a _____ (what kind) fire detection system.

7. The reference junction in a thermocouple fire-detection system is mounted between _____ blocks so its temperature will not rise as rapidly as the temperature of the hot junctions.

8. Current from the thermocouples closes a _____ relay which connects the slave relay to the battery to signal a fire.

FIRE PROTECTION SYSTEMS — CHAPTER TEN

9. The Kidde continuous-loop fire-detection system uses, as a detector, two wires imbedded in a ceramic material that changes its _____ as its temperature changes.

10. A Fenwal continuous-loop fire-detection system uses a single wire, insulated from its housing by ceramic beads covered with a eutectic salt whose electrical _____ decreases drastically as the temperature of the beads rises to the warning temperature.

11. A Kidde continuous-loop fire-detection system _____ (will or will not) warn of a general overheat condition as well as a fire.

12. The sensitive element in a Lindberg fire-detection system is sealed inside a stainless steel tube. When this element gets hot, it releases a _____ . A pressure switch built into the tube closes when the pressure reaches a sufficiently high value, and it signals the presence of a fire or overheat condition.

13. A fire that has a flammable petroleum product as its fuel is a Class _____ fire.

14. A fire that has paper as its fuel is a Class _____ fire.

15. A fire that involves energized electrical equipment is a Class _____ fire.

16. Fire zones are also classified with class letters. Identify the class of fire zone that each of these locations represents:

   A. The engine accessory compartment of a radial engine is a Class _____ fire zone.

   B. A wheel well is a Class _____ fire zone.

   C. The power section of a reciprocating engine is a Class _____ fire zone.

   D. Heat exchanger ducts in a reciprocating engine installation are Class _____ fire zones.

CHAPTER TEN   FIRE PROTECTION SYSTEMS

17. Three fire-extinguishing agents best for extinguishing aircraft fires are

    A. _____

    B. _____

    C. _____

18. Carbon tetrachloride _____ (is or is not) a satisfactory agent for extinguishing aircraft fires.

19. CB is used as an extinguishing agent in some noninstalled fire-extinguishing systems. The main reason it is not good for aircraft fires is that it is _____ .

20. Most modern jet aircraft are protected by high-rate-discharge bottles containing an agent such as Halon 1211 or Halon 1301 pressurized with _____ .

21. Built-in fire-extinguisher systems installed in some of the older aircraft use _____ as the extinguishing agent.

22. Pulling the fire-pull T-handle in a jet transport aircraft _____ (does or does not) discharge the HRD bottles into the fire.

23. The fire-extinguishing agent is released from an HRD bottle by a cutter that is driven into the bottle seal by an _____ .

24. When a fire-extinguishing system has been discharged normally, a _____ (what color) disk is blown out of the discharge indicator.

25. When a fire-extinguishing system has been discharged because of an overheat condition, a _____ (what color) disk is blown out of the discharge indicator.

26. The status of charge of a fire-extinguisher HRD bottle is determined by the _____ built into the bottle.

27. The discharge cartridge installed in an HRD bottle should be replaced after a given number of _____ .

AIRFRAME SECTION WORKBOOK

FIRE PROTECTION SYSTEMS — CHAPTER TEN

28. The state of charge of a hand-held carbon dioxide fire extinguisher is determined by its _____.

29. Two types of fire extinguishers that are suitable for aircraft cockpit and cabin fires are

   A. _____

   B. _____

30. Two types of smoke-detector systems that are installed on aircraft are

   A. _____

   B. _____

31. A carbon monoxide detector is a small button of indicating agent that changes its _____ in the presence of carbon monoxide.

32. As little as _____ (how many) parts of carbon monoxide per million parts of air will cause unconsciousness in an hour and death in two to three hours.

CHAPTER 11

AIRCRAFT ELECTRICAL SYSTEMS

1. Two or more separately insulated conductors in the same jacket make up what is known as a _____ cable.

2. Aircraft electrical wiring is measured by the _____-wire gage.

3. When using a round-wire gage, the wire size is measured by the _____ (slot or circular opening) in the gage.

4. According to the American wire gage, the smaller the gage number, the _____ (larger or smaller) the wire diameter.

5. Three things that must be taken into consideration when choosing a wire size for installation in an aircraft are

   A. _____

   B. _____

   C. _____

6. The two most widely used materials for aircraft electrical wiring are

   A. _____

   B. _____

7. The main advantage of aluminum for aircraft wiring is its _____ .

8. Using the table in Fig. 11-2, on page 434 of the text, find the resistance of 100 feet of 20-gage solid copper wire at a temperature of 25 degrees Celsius. The resistance is _____ ohms.

AIRCRAFT ELECTRICAL SYSTEMS CHAPTER ELEVEN

9. According to the chart in Fig. 11-5, on page 436 of the text, the maximum amount of current that should be carried in a 10-gage wire insulated with a thermoplastic insulation is _____ amps.

10. According to the chart in Fig. 11-6, on page 436 of the text, the maximum voltage drop allowed in a 28-volt system for a continuous load is _____ volt/volts.

11. According to the chart in Fig. 11-6, on page 436 of the text, the maximum voltage drop allowed in a 14-volt system for an intermittent load is _____ volt/volts.

12. In a single-wire aircraft electrical system, the return current from a component flows through the _____ .

13. Using the wire chart in Fig. 11-7, on page 437 of the text, find the size wire you would need to supply a 30-amp continuous load in a 28-volt electrical system if the load is 60 feet from the bus and the wire is to be routed in a bundle. You should use a _____-gage wire.

14. Using the wire chart in Fig. 11-7, on page 437 of the text, find the smallest size wire you could use to supply a 50-amp continuous load in a 28-volt electrical system if the load is 16 feet from the bus and the wire is not routed in a bundle. You could use a _____-gage wire.

15. Using the wire chart in Fig. 11-8, on page 438 of the text, find the size wire you would need to supply a 100-amp intermittent load in a 14-volt electrical system. The load is 20 feet from the bus, and the wire is not routed in a bundle. You should use a _____-gage wire.

16. An intermittent electrical load is considered to be one that draws current for a maximum of _____ minutes.

17. Two factors to consider when selecting the type of insulation for an electrical wire are

   A. _____

   B. _____

18. Aircraft electrical wires should be marked with an identification number at intervals along their length and at a point within _____ inches of each junction or terminal block.

CHAPTER ELEVEN — AIRCRAFT ELECTRICAL SYSTEMS

19. Wiring not installed in a conduit is called _____ wiring.

20. For practical installation purposes, wire bundles should be limited to contain no more than _____ wires.

21. Magnetic fields surrounding wires routed near a magnetic compass may be cancelled by _____ the wires.

22. Noninsulated splices in a wire bundle should be covered with _____ tubing tied at both ends.

23. The maximum amount of slack allowed in a wire bundle between supports is that which will allow the bundle to be deflected _____ inch with normal hand force.

24. Wire bundles should be routed so they will not be bent with a radius of less than _____ times the outside diameter of the bundle.

25. For the neatest routing in an aircraft structure, wire bundles should be parallel or at right angles to the _____ or _____ in the area through which they are routed.

26. Wire bundles are supported at each point they pass through a bulkhead with a _____ clamp.

27. When a wire bundle comes to within 1/4-inch of the edge of the hole through which it passes, a _____ must be installed in the hole.

28. Wires that must be routed in an area of high temperature should be insulated with _____, _____, or _____.

29. Wire bundles should not be routed in the lowest _____ inches of any structure that might collect any fluids.

30. Plastic tubing used to protect wire bundles should have a _____-inch-diameter drain hole at the lowest point in the tubing.

AIRCRAFT ELECTRICAL SYSTEMS CHAPTER ELEVEN

31. Wire bundles routed through wheel wells should be protected by enclosing them in _____ .

32. If a wire bundle must be routed parallel to a fluid line, the wire should be _____ (above or below) the fluid line.

33. The minimum separation allowed between a fluid line and a wire bundle is _____ inch.

34. Cable clamps used to secure a wire bundle to an aircraft structure should have _____ in them to protect the wires.

35. If cotton cord is used for lacing or tieing a wire bundle, the cord should be _____ before it is used.

36. Single-cord lacing of a wire bundle is started with a _____ hitch and an extra loop.

37. Double-cord lacing of a wire bundle is started with a _____ knot.

38. Individual spot ties may be used on a wire bundle instead of lacing the bundle. The spot ties are made with a _____ hitch and a square knot.

39. The preferred device for splicing aircraft wires is a _____ splice connector.

40. Terminals to be swaged onto aluminum wire should be made of _____ .

41. Terminals to be swaged onto aluminum wire should be filled with a petrolatum-_____ compound to abrade the oxide off of the wire and to minimize later oxidation.

42. When attaching an aluminum wire terminal to an aircraft structure, the terminal should be protected from corrosion by installing _____ (what material) plain washers on both sides of the terminal.

CHAPTER ELEVEN — AIRCRAFT ELECTRICAL SYSTEMS

43. Most bonding jumpers attached to an aircraft structure are made of _____.

44. The maximum number of wires that should be connected to any single stud in a terminal block is _____.

45. A conduit used to protect a wire bundle should have an inside diameter _____ percent larger than the diameter of the wire bundle.

46. An ammeter installed in an alternator lead should be red-lined at _____ percent of the alternator rating.

47. If it is not practical for the flight crew to monitor the electrical load in an aircraft, the maximum continuous connected load should be no more than _____ percent of the generator rated output capacity.

48. Fuses and circuit breakers are installed in an aircraft electrical system to protect the _____ (wiring or components).

49. A 16-gage wire should be protected with a _____-amp circuit breaker.

50. A circuit breaker that will open a circuit if a fault occurs regardless of the position of the operating control is called a _____ circuit breaker.

51. A switch that controls an incandescent lamp load of 10 amps in a 24-volt circuit should have a rating of at least _____ amps.

52. It is generally recommended that the operating control of a switch be in the up position when the switch is _____ (on or off).

53. The green navigation light for an airplane is mounted on the tip of the _____ (right or left) wing.

54. Three types of faults that cause most of the trouble in aircraft electrical systems are

   A. _____

   B. _____

   C. _____

AIRFRAME SECTION WORKBOOK

AIRCRAFT ELECTRICAL SYSTEMS CHAPTER ELEVEN

55. Before using an ohmmeter to troubleshoot an electrical system, the part of the system to be checked should be _____ (connected to or disconnected from) the source of voltage.

CHAPTER 12

AIRCRAFT INSTRUMENT SYSTEMS

1. Two ways of grouping aircraft instruments are

   A. _____

   B. _____

2. Three types of instruments based on their function are

   A. _____

   B. _____

   C. _____

3. Magnetic fields produced by some of the electrical instruments in an aircraft are prevented from interfering with adjacent equipment by mounting the instrument in a _____ case.

4. The location of range marks required on an aircraft instrument may be found in the _____ for the aircraft.

5. The normal operating range of the condition being measured by an instrument is shown on the dial by a _____ (what color) arc.

6. A caution range of the condition being measured by an instrument is shown on the dial by a _____ (what color) arc.

7. The never-exceed limit of the condition being measured by an instrument is shown on the dial by a _____ (what color) radial line.

8. The range of operation permitted only under certain conditions is shown on an instrument dial by a _____ (what color) arc.

AIRCRAFT INSTRUMENT SYSTEMS   CHAPTER TWELVE

9. If range marks are placed on the instrument glass rather than the dial, a slip mark should be made between the glass and the instrument bezel. This mark is _____ (what color).

10. Instruments that do not have a mounting flange on their case are held in the instrument panel by a _____ that is fastened to the back of the panel.

11. Instrument panels are shock mounted to absorb _____ (low or high)-frequency vibrations.

12. Engine oil pressure is usually measured by an instrument using a _____ tube mechanism.

13. An engine gage unit is actually three instruments in one case. The instruments are normally

   A. _____

   B. _____

   C. _____

14. Instruments used for measuring low values of pressure often use a _____ rather than a bourdon tube.

15. Differential pressure is normally measured with a _____-type pressure gage.

16. The manifold pressure gage used with a reciprocating engine measures _____ (absolute, differential, or gage) pressure.

17. When the engine is not running, a manifold pressure gage reads the pressure of the surrounding _____.

18. An airspeed indicator measures _____ (absolute, differential, or gage) pressure.

CHAPTER TWELVE — AIRCRAFT INSTRUMENT SYSTEMS

19. Three flight instruments that sense the pressure at the aircraft static port are

    A. _____

    B. _____

    C. _____

20. The airspeed indicator senses pressure taken from two locations; these are the

    A. _____

    B. _____

21. A preflight check of the pitot heater is made by turning the heater on and watching the _____ for an indication of current flow.

22. Most aircraft that operate into known icing conditions have an alternate source for _____ (static or pitot) air pressure.

23. An altimeter measures _____ (absolute, differential, or gage) pressure.

24. A rate of climb indicator measures the rate at which the _____ changes when the aircraft goes up or down.

25. An instantaneous-rate-of-climb indicator eliminates the lag in the indication by using air pumps operated by _____ .

26. The airspeed range in which flaps can be used on an airplane is shown on an airspeed indicator by a _____ (what color) arc.

27. The maximum-speed pointer on a maximum-allowable airspeed indicator is operated by an _____ mechanism.

28. Three different things that are sensed by a true-airspeed indicator to drive a single pointer over the dial are

    A. _____

    B. _____

    C. _____

AIRCRAFT INSTRUMENT SYSTEMS — CHAPTER TWELVE

29. A Mach indicator shows the ratio between the airspeed of the airplane and the _____ under the existing atmospheric conditions.

30. Two other names for a turn and bank indicator are

    A. _____

    B. _____

31. An aircraft using air-driven attitude gyro instruments normally uses an _____-operated turn and slip indicator.

32. A standard-rate turn as shown on a turn and slip indicator is _____ (how many) degrees per second.

33. The transmitter used with a DC selsyn remote indicating system is a variable _____.

34. The rotor in a Magnesyn remote indicating system is a _____ (permanent magnet or electromagnet).

35. The tank unit in a capacitor-type fuel-quantity-indicating system is a cylindrical _____.

36. The dielectric constant of air is _____, and the dielectric constant of aircraft fuel is approximately _____.

37. Variations in the actual dielectric constant of the fuel is compensated for by a _____ capacitor that is part of one of the tank probes.

38. The angle-of-attack indicator described on page 488 of the text measures the angle of attack by a differential pressure which changes as the direction the air flowing across the slots in the _____ detector probe changes.

39. The tachometer used on a reciprocating engine indicates the engine speed in _____.

CHAPTER TWELVE AIRCRAFT INSTRUMENT SYSTEMS

40. The tachometer used with a turbine engine indicates the engine speed in _____ of the rated engine speed.

41. The transmitter for an electric tachometer is _____-phase AC generator.

42. The pointer in an electric tachometer indicator is driven by a _____ motor.

43. The propeller-shaped pointer on a synchroscope dial rotates at a speed that is proportional to the _____ in speed of the engines.

44. The pickup probe for an electrical oil-temperature indicator is a coil of nickel/manganese wire whose _____ changes as its temperature changes.

45. Three combinations of wires used as the thermocouples for temperature-measuring instruments are

   A. _____
   B. _____
   C. _____

46. The exhaust gas temperature measuring system installed on a gas turbine engine uses a number of thermocouples installed in the tailpipe of the engine. These thermocouples are connected together in _____ (series or parallel).

47. A vane-type fuel flowmeter measures the _____ (mass or volume) of fuel flowing to the engine.

48. A turbine-type fuel flowmeter measures the _____ (mass or volume) of fuel flowing to the engine.

49. The two basic characteristics that make a gyroscope a useful measuring device are

   A. _____
   B. _____

AIRFRAME SECTION WORKBOOK

AIRCRAFT INSTRUMENT SYSTEMS — CHAPTER TWELVE

50. When a gyroscopic instrument is connected to the suction side of a vacuum pump, the air pressure used to spin the rotor comes from the _____.

51. The air pump used in most reciprocating-engine-powered aircraft to drive the gyroscopic instruments is a _____ pump.

52. The amount of suction specified for driving gyroscopic instruments is measured in _____.

53. An attitude gyro installed in an aircraft shows the pilot the attitude of the aircraft relative to two axes; these are

   A. _____

   B. _____

54. The gyroscopic characteristic used to keep an attitude gyro erect is called _____.

55. The magnetic compass error that is minimized by "swinging" the compass is _____.

56. The system in an autopilot that smoothly returns the flight surfaces to their streamlined position when the required changes in flight attitude have been made is called the _____ system.

57. A flight instrument system that shows the pilot the direction and amount the attitude of the aircraft must be changed to achieve a desired condition is called a _____ _____ system.

58. A panel on an aircraft that contains all of the indicator lights to show the conditions of the various systems is called an _____ panel.

CHAPTER 13

COMMUNICATIONS AND NAVIGATION SYSTEMS

1. An A & P mechanic license _____ (does or does not) authorize the holder to repair communications and navigation equipment.

2. The license required for a person to calibrate radio transmitting equipment is issued by the _____ .

3. Radio communications is based on the transmission of electrical energy through space in the form of _____ energy waves.

4. The conductor in a radio transmitting system that radiates the electromagnetic waves is called the _____ .

5. Radio waves travel at the speed of _____ .

6. When an electromagnetic wave cuts across a receiving antenna, _____ are forced to move back and forth in the antenna.

7. VHF radio equipment used in aircraft operates in the frequency range between _____ and _____ megahertz.

8. The high-frequency alternating current used in a radio transmitter is produced by an _____ .

9. The circuit used in a radio transmitter to increase the amount of energy in the signal is called an _____ .

10. The circuit used in a radio transmitter to put the intelligence on the RF carrier signal is called a _____ .

COMMUNICATIONS AND NAVIGATION SYSTEMS — CHAPTER THIRTEEN

11. The two types of modulation used in aircraft radio equipment are

    A. _____

    B. _____

12. The frequency of the carrier wave produced in an aircraft radio transmitter is controlled by a _____.

13. The removal of intelligence from a radio wave is done in the receiver by a circuit called a _____.

14. The demodulator circuit used in an FM radio receiver is called a _____.

15. A device used in an aircraft to change DC from the battery into AC that can be used in the radio equipment is called an _____.

16. A piece of radio equipment that contains both a transmitter and a receiver in one housing is called a _____.

17. VHF radio signals _____ (do or do not) follow the curvature of the earth.

18. Any person operating an aircraft radio transmitter must hold a _____ permit issued by the FCC.

19. Any aircraft in which an operating radio transmitter is installed must display a radio _____ license issued by the FCC.

20. Long-range communications is done with aircraft radio equipment that operates in the _____ (HF, VHF, or UHF) range.

21. The most widely used radio navigation system in the United States is the VHF _____ system.

22. The course to be flown by an airplane using VOR for navigation is shown on the _____ indicator.

CHAPTER THIRTEEN — COMMUNICATIONS AND NAVIGATION SYSTEMS

23. A VOR system follows a line that extends out from the transmitter like the spoke of a wheel. Each line is called a _____.

24. The pilot selects the radial he wishes to fly along by rotating the _____ in the CDI.

25. A mechanic can check the accuracy of VOR equipment by tuning to a _____ signal which can be received on most airports.

26. An unusable VOR signal is indicated by a _____ that shows in the instrument.

27. Three components of an instrument landing system (ILS) installed in an aircraft are

   A. _____
   B. _____
   C. _____

28. The component in an ILS that shares the antenna with VOR is the _____.

29. The glide slope portion of an ILS operates in the _____ (UHF or VHF) frequency band.

30. Glide slope information is shown to the pilot by the _____ (horizontal or vertical) needle in the CDI used by the VOR.

31. The marker beacon receiver is tuned to a single fixed frequency of _____ megahertz.

32. Information from a marker beacon receiver is shown to the pilot by the illumination of one of _____ (how many) colored lights.

33. Distance measuring equipment (DME) operates in the _____ (UHF or VHF) frequency band.

34. The DME antenna is a short stub that is mounted on the _____ (top

COMMUNICATIONS AND NAVIGATION SYSTEMS

or bottom) of an aircraft fuselage.

35. The automatic direction finder (ADF) operates in the _____ (LF, HF, or VHF) frequency band and in the commercial broadcast band.

36. An ADF system requires two antennas; these are

   A. _____

   B. _____

37. The electronic system in an aircraft that provides a signal to the air traffic controller on the ground to identify the aircraft is the _____.

38. Doppler navigation systems _____ (do or do not) require the use of a ground station.

39. The most accurate navigation system used on airliners and military aircraft is the

   _____.

40. An airborne system that transmits a short, high-energy pulse of electromagnetic radiation and receives a return of this energy after it bounces off of something is called a _____ system.

41. A radio altimeter gives the pilot of an aircraft a direct measure of the distance between the aircraft and the _____.

42. An emergency locator transmitter (ELT) transmits a signal on two frequencies; these are

   A. _____

   B. _____

43. The batteries in an ELT must be replaced after a specified time. The date of required replacement must be displayed on the _____.

44. When installing a piece of electronic equipment in a shock-mounted rack, you must be sure that the _____ jumper is sufficiently large to carry all of the return current from the equipment.

CHAPTER THIRTEEN — COMMUNICATIONS AND NAVIGATION SYSTEMS

45. When all of the wiring and components in an electronic system are enclosed in conductive housings or in metallic braid, the system is said to be _____.

46. Electrostatic energy that collects on aircraft control surfaces as the aircraft flies through the air is dissipated into the air through _____.

47. When installing a radio antenna on the outside of an aircraft skin, you should normally reinforce the skin by installing a _____ on the inside of the skin.

48. A radio transmitter is normally connected to its antenna with _____ cable.

CHAPTER 14

CABIN ATMOSPHERE CONTROL SYSTEMS

1. Reduced efficiency of a flight crew caused by a lack of oxygen in the air that is breathed is called _____.

2. The first indication of hypoxia is _____ and fatigue.

3. The atmosphere that surrounds the earth is made up basically of two gases; these are

   A. _____

   B. _____

4. The standard pressure of the atmosphere at the surface of the Earth is

   A. _____ pounds per square inch

   B. _____ inches of mercury

   C. _____ millibars

5. The temperature of the atmosphere decreases with an increase in altitude until the _____ is reached.

6. The lowest layer of our atmosphere is called the _____.

7. The layer of our atmosphere in which the temperature remains constant is called the _____.

8. The use of supplemental oxygen _____ (is or is not) required for the occupants of a pressurized aircraft.

9. The amount of pressure inside the cabin of a pressurized aircraft is controlled by the operation of _____ valves.

10. The amount an aircraft cabin can be pressurized is determined by the _____ of the aircraft.

AIRCRAFT ATMOSPHERE CONTROL SYSTEMS   CHAPTER FOURTEEN

11. Pressure that is referenced from a vacuum is called _____ pressure.

12. A change in conditions that is accomplished with no transfer of heat energy is called an _____ change.

13. Air for pressurizing the cabin of a turbine-engine-powered aircraft is normally taken from an _____ .

14. Two types of cabin superchargers that can be used on a reciprocating-engine-powered aircraft are

    A. _____

    B. _____

15. Two modes of operation of a cabin supercharging system are

    A. _____

    B. _____

16. The mode of operation of a cabin supercharger system that maintains a constant cabin altitude as the flight altitude changes is called the _____ mode.

17. The mode of operation of a cabin supercharger system that maintains a constant pressure differential between the inside of the cabin and the outside air is called the _____ mode.

18. The component in a cabin supercharger system that prevents cabin pressure exceeding a predetermined pressure above the outside air pressure is the _____ valve.

19. Combustion heaters used in some aircraft get their fuel from the aircraft _____ .

20. The cabin temperature maintained by a combustion heater system is controlled by the thermostat which controls the heater _____ .

CHAPTER FOURTEEN — AIRCRAFT ATMOSPHERE CONTROL SYSTEMS

21. Two types of air cooling systems used in aircraft are

    A. _____

    B. _____

22. The cold air produced in an air-cycle cooling system is obtained by extracting energy from the pressurizing air by passing it through an _____ turbine.

23. Temperature of the cabin air is controlled in an air-cycle cooling system by mixing cold air from the refrigeration unit with hot _____ air.

24. Water that condenses from the air after it passes through the expansion turbine is removed by the _____.

25. In a vapor-cycle air-conditioning system, heat is picked up from the cabin as the refrigerant flows through the _____.

26. Heat is carried out of the aircraft cabin and discharged into the outside air by the refrigerant as it passes through the _____.

27. The refrigerant in a vapor-cycle air-conditioning system changes from a vapor into a liquid in the _____.

28. The refrigerant in a vapor-cycle air-conditioning system changes from a liquid into a vapor in the _____.

29. The unit in a vapor-cycle air-conditioning system that raises the pressure of the refrigerant gas is the _____.

30. The unit in a vapor-cycle air-conditioning system that drops the pressure of the liquid refrigerant is the _____.

31. The most generally used refrigerant for aircraft air-conditioning systems is refrigerant _____ (12 or 22).

AIRCRAFT ATMOSPHERE CONTROL SYSTEMS — CHAPTER FOURTEEN

32. Two types of oxygen systems used in aircraft are

   A. _____

   B. _____

33. Hospital oxygen _____ (is or is not) approved for use in an aircraft oxygen system.

34. Solid oxygen systems release their oxygen when the sodium chlorate candle is _____.

35. Demand oxygen regulators allow oxygen to flow into the mask only when the wearer of the mask _____.

36. The diluter-demand oxygen regulator dilutes the oxygen with _____.

37. The amount of oxygen in an oxygen cylinder is indicated by the _____ of the oxygen.

38. Oxygen systems are checked for leaks by checking for bubbles in a film of a special soap solution made with _____ soap.

ANSWERS

AIRFRAME SECTION WORKBOOK ANSWERS

CHAPTER 1

1. A. Fuselage
   B. Wings
   C. Stabilizers
   D. Flight control surfaces
   E. Landing gear
2. A. Fuselage
   B. Main rotor and related gearbox
   C. Tail rotor
   D. Landing gear
3. A. Tension
   B. Compression
   C. Torsion
   D. Shear
   E. Bending
4. stress
5. strain
6. tension
7. compression
8. torsion
9. shear
10. bending
11. A. Truss type
    B. Monocoque type
12. A. Monocoque
    B. Semimonocoque
    C. Reinforced shell
13. skin
14. semimonocoque
15. longerons
16. stringers
17. bulkheads
18. gusset
19. stations
20. datum
21. inches
22. buttock
23. water
24. cantilever
25. spar
26. zero
27. ribs
28. laminated
29. extruded
30. caps
31. web
32. false
33. drag
34. anti-drag
35. compression
36. wet
37. fiberglass or aluminum
38. nacelle, pod
39. empennage
40. ailerons
41. rudder

AIRFRAME SECTION WORKBOOK

ANSWERS

42. elevators
43. A. Trailing-edge flaps
    B. Leading-edge flaps
    C. Spoilers
    D. Leading-edge slats
44. behind
45. balance panels
46. lift
47. area
48. camber
49. longitudinal
50. riveting
51. fairing

CHAPTER 2

1. assembly
2. rigging
3. Type Certificate Data Sheets
4. nitrogen and oxygen
5. 14.7
6. 29.92
7. increases
8. decreases
9. faster
10. less
11. decrease
12. A. Lift
    B. Drag
    C. Thrust
    D. Gravity
13. velocity
14. acceleration
15. external force
16. mass

17. A. Shape of the airfoil section
    B. Angle of attack
    C. Speed of the air over the surface
    D. Area of the wing
    E. Density of the air
18. attack
19. pressure
20. forward
21. critical
22. incidence
23. fineness
24. drag
25. camber
26. aspect
27. center of pressure
28. ahead
29. perpendicular
30. parallel
31. A. Parasite drag
    B. Profile drag
    C. Induced drag
32. vortices
33. longitudinal
34. lateral
35. vertical
36. roll
37. pitch
38. yaw
39. stability
40. maneuverability
41. controllability
42. static
43. negative
44. horizontal stabilizer
45. more

ANSWERS

46. lateral
47. increased
48. smaller
49. upward
50. before
51. skidding
52. stabilator
53. ruddervators
54. trim
55. opposite
56. opposite
57. spring
58. A. Camber
    B. Area
59. boundary
60. tip-path plane
61. torque
62. anti-torque
63. precession
64. advancing
65. retreating
66. dissymmetry
67. dissymmetry
68. drag
69. articulated
70. centrifugal
71. one-half
72. autorotative
73. A. Longitudinal
    B. Lateral
    C. Vertical
74. collective
75. cyclic
76. collective
77. temperature
78. subsonic
79. transonic
80. supersonic
81. supersonic
82. above
83. below
84. speeds up
85. vortex
86. normal
87. low speed
88. irreversible
89. feel
90. least
91. A. Cable system
    B. Push-pull system
    C. Torque-tube system
92. replaced
93. turnbuckles
94. are
95. three
96. air seals
97. arm
98. before
99. temperature
100. 140
101. propeller
102. spirit
103. loose
104. reflectors
105. behind

CHAPTER 3

1. A. Cotton
   B. Linen

ANSWERS

2. A. Polyamide (Nylon)
   B. Acrylic (Orlon)
   C. Polyester (Dacron)
3. never-exceed, wing
4. 80
5. 80 and 84
6. stronger
7. may
8. 4.5
9. second
10. three
11. reinforcing
12. 14
13. beeswax
14. two
15. celluloid
16. A. French fell
    B. Folded fell
17. 8 and 10
18. 1/2
19. six
20. four
21. four
22. 200
23. should not
24. four, four
25. four, three
26. aluminum
27. water
28. reinforcing
29. clear dope
30. A. The envelope method
    B. The blanket method
31. chordwise
32. nitrate
33. water
34. heat
35. two
36. 225
37. nonshrinking
38. trailing
39. slipstream
40. 2-1/2, 3-1/2
41. one
42. splice
43. 1/2
44. seine
45. should not
46. behind
47. baseball
48. 1/4, 1/4
49. 8, 10
50. 1-1/2
51. clear
52. 16
53. should not
54. three
55. 150, 16
56. two
57. front spar
58. three
59. four, eight
60. after
61. first
62. brushing
63. aluminum
64. dark
65. is not
66. 70
67. airworthy

ANSWERS

68. second
69. static
70. blush
71. A. Cellulose-nitrate
    B. Cellulose-acetate-butyrate
72. nitrate
73. 1-3/4

CHAPTER 4

1. quickly
2. denatured
3. zinc chromate
4. A. Film-base compound
    B. Plasticizer
    C. Solvents
    D. Diluents
    E. Slow-dryers
    F. Pigments
5. can
6. two
7. is not
8. spar varnish
9. swell
10. sandpaper
11. polyurethane
12. converter, resin
13. 15
14. long
15. six
16. cup
17. parallel
18. A. Paper
    B. Metal
    C. Vinyl film

CHAPTER 5

1. manufacturer
2. strength
3. 1/2
4. is not
5. compressive
6. inboard
7. 5/32
8. 19
9. negligible
10. A. Tension
    B. Compression
    C. Torsion
    D. Shear
    E. Bending
11. compression
12. tension
13. shear
14. A. Tension
    B. Compression
15. Cleco
16. can
17. 3/16
18. pneumatically
19. aluminum oxide
20. thin
21. cornice
22. slip roll
23. drop hammers
24. is not
25. sharp lead pencil
26. bend allowance
27. A. The degree of bend
    B. The radius of the bend
    C. The thickness of the metal

ANSWERS

    D. The type of metal used
28. neutral
29. inside
30. 0.432
31. 0.421
32. mold
33. setback
34. bend radius
35. 0.70
36. sight
37. closed
38. open
39. 3/8
40. 1/2
41. 5.768
42. relief
43. bend radius
44. inside
45. flanged
46. ends
47. center
48. around the edges
49. joggle
50. flatter
51. slower
52. more
53. dry powder
54. thicker
55. 1-1/2
56. 1-1/2
57. center
58. two
59. 2117-T
60. pitch
61. three
62. transverse pitch
63. 75
64. countersink
65. dimpling
66. squeeze
67. 30
68. dimpled
69. A. Radius dimpling
    B. Coin dimpling
70. hot
71. three
72. three and four
73. 1/2
74. Microshaver
75. shear
76. bearing
77. 31
78. can
79. cannot
80. Rivnut
81. keyed
82. 75
83. heavyweight
84. collar
85. octagonal
86. round
87. three
88. four
89. corrosion
90. 18
91. delamination
92. ring
93. router
94. Methyl-Ethyl-Ketone (MEK)
95. glass fabric

ANSWERS

96. one
97. micro-balloons
98. A. Thermoplastic
    B. Thermosetting
99. thermoplastic
100. thermosetting
101. acetate
102. aliphatic
103. crazing
104. flatter
105. soak
106. below
107. soap
108. polyethylene
109. electrical
110. spruce
111. cannot
112. synthetic resin
113. 125 and 150
114. 10
115. should not
116. scarf
117. five
118. 12
119. one
120. may not

CHAPTER 6

1. A. Gas welding
   B. Electric arc welding
   C. Electric resistance welding
2. A. Acetylene
   B. Hydrogen
3. oxyhydrogen
4. A. Metallic-arc welding
   B. Inert-gas welding
5. A. Butt welding
   B. Spot welding
   C. Seam welding
6. 15
7. acetone
8. 1,800
9. beeswax
10. A. Balanced-pressure torch
    B. Injector-type torch
11. balanced-pressure
12. heat
13. 5,700 to 6,300
14. neutral
15. reducing
16. oxidizing
17. backfire
18. flashback
19. acetylene
20. 1/8
21. forehand
22. stop drilled
23. ammonium nitrate
24. neutral
25. is not
26. carburizing (reducing)
27. is
28. sulfuric
29. does not
30. neutral
31. is not
32. oxygen
33. is not
34. capillary

ANSWERS

35. silver soldering
36. lead and tin
37. solder
38. 10,000
39. consumable
40. is not
41. nonconsumable
42. A. Argon
    B. Helium
43. consumable
44. plasma
45. should not
46. inner
47. outer
48. 1/10, 1/4, 1
49. 1/8

CHAPTER 7

1. A. Rime ice
   B. Glaze ice
2. A. There must be visible moisture in the air.
   B. Temperature must be near or below freezing.
3. can
4. anti-icing
5. deicing
6. vacuum
7. Rivnuts
8. cement
9. vacuum
10. vane
11. wet
12. carbon
13. oil separator
14. timer

15. mild soap and water
16. combustion heaters
17. augmentor
18. compressor
19. expansion bellows
20. ethylene glycol and isopropyl alcohol
21. hot air
22. conductive
23. AC
24. should not
25. alcohol
26. electric
27. A. Blowing it off with high-velocity air
    B. Wiping it off with a rubber-blade windshield wiper
    C. Using chemical rain repellant
28. should not

CHAPTER 8

1. A. Vegetable-base fluid
   B. Mineral-base fluid
   C. Phosphate ester-base fluid
2. vegetable
3. castor, alcohol
4. red
5. fire resistant
6. phosphate ester
7. paper
8. 1/25,400
9. bypass relief
10. A. Reservoir
    B. Pump
    C. Selector valve
    D. Actuating unit

ANSWERS

11. accumulator
12. check
13. pressurized
14. A. Air brought in through a venturi tee
    B. Compressor bleed air
    C. Hydraulic fluid pressure acting on a bellowfram
15. normal
16. double-action
17. A. Constant-delivery pump
    B. Variable-delivery pump
18. around
19. five
20. high, low
21. high
22. shear
23. constant
24. would
25. unloading
26. A. Diaphragm-type
    B. Bladder-type
    C. Piston-type
27. nitrogen
28. orifice
29. pump
30. single-action
31. balanced
32. A. Rotary selector valves
    B. Spool-type selector valves
33. compressed air
34. electric or hydraulic
35. bleed air
36. vane
37. restrictor
38. shuttle
39. hydraulic
40. heater

CHAPTER 9

1. bogie
2. oleo
3. metering
4. snubbing
5. locating
6. air valve
7. nitrogen
8. strut
9. bleeding
10. torque links
11. trunnion
12. drag strut
13. drag
14. sequence
15. torque links
16. red streamer
17. green
18. steering
19. shimmy damper
20. rapid
21. steer
22. A. Independent brake systems
    B. Power brake systems
    C. Power-boost brake systems
23. compensating
24. accumulator
25. regulator
26. debooster
27. decreases
28. power boost
29. 15

ANSWERS

30. single-disk
31. dual-disk
32. segmented rotor
33. stator
34. neoprene
35. A. Gravity bleeding
    B. Pressure bleeding
36. spongy
37. split
38. rubber packing
39. thermal
40. tapered roller
41. more
42. overinflated
43. underinflated
44. two
45. five to ten
46. five
47. 48
48. is
49. is
50. should
51. vegetable soap
52. heavy
53. light
54. deflated
55. before
56. across the wheel from
57. dial
58. original cartons
59. axle
60. skid control
61. 15 to 20
62. is

CHAPTER 10

1. zone
2. A. Overheat detector
   B. Rate-of-temperature-rise detector
   C. Flame detector
3. will
4. will
5. will not
6. rate-of-temperature-rise
7. insulation
8. sensitive
9. resistance
10. resistance
11. will
12. gas
13. B
14. A
15. C
16. A. C
    B. D
    C. A
    D. B
17. A. Halon 1301
    B. Halon 1211
    C. Carbon dioxide
18. is not
19. corrosive to aluminum
20. nitrogen
21. carbon dioxide
22. does not
23. explosive charge
24. yellow
25. red
26. pressure gage
27. hours

28. weight
29. A. Halon 1301
    B. Carbon dioxide
30. A. Photoelectric
    B. Visual
31. color
32. 800

CHAPTER 11

1. multiconductor
2. American
3. slot
4. larger
5. A. Allowable power loss
   B. Permissible voltage drop
   C. Current-carrying capability
6. A. Copper
   B. Aluminum
7. low weight
8. 1.04
9. 55
10. one
11. one
12. aircraft structure
13. six
14. 10
15. six
16. two
17. A. Insulation resistance
    B. Dielectric strength
18. three
19. open
20. 75
21. twisting

22. plastic
23. 1/2
24. 10
25. stringers or ribs
26. cable
27. grommet
28. asbestos, fiberglass, Teflon
29. four
30. 1/8
31. flexible tubing
32. above
33. 1/2
34. cushions
35. waxed
36. clove
37. bowline-on-a-bight
38. clove
39. self-insulating
40. aluminum
41. zinc dust
42. aluminum
43. aluminum alloy
44. four
45. 25
46. 100
47. 80
48. wiring
49. 15
50. trip-free
51. 80
52. on
53. right
54. A. Open circuit
    B. Short circuit
    C. Low-power circuit

ANSWERS

55. disconnected from

CHAPTER 12

1. A. The job they perform
   B. The principle on which they work
2. A. Flight instruments
   B. Engine instruments
   C. Navigation instruments
3. steel
4. Type Certificate Data Sheets
5. green
6. yellow
7. red
8. blue
9. white
10. clamp
11. low
12. bourdon
13. A. Oil pressure
    B. Oil temperature
    C. Fuel pressure
14. diaphragm
15. diaphragm
16. absolute
17. atmosphere
18. differential
19. A. Airspeed indicator
    B. Altimeter
    C. Rate-of-climb indicator
20. A. Pitot tube
    B. Static port
21. ammeter
22. static
23. absolute
24. pressure
25. accelerometers
26. white
27. altimeter
28. A. Pitot pressure
    B. Static pressure
    C. Free-air temperature
29. speed of sound
30. A. Turn and slip indicator
    B. Needle and ball indicator
31. electrically
32. three
33. resistor
34. permanent
35. capacitor
36. one, two
37. compensating
38. airstream direction
39. hundreds of RPM
40. percent
41. three
42. synchronous
43. difference
44. resistance
45. A. Iron-constantan
    B. Copper-constantan
    C. Chromel-alumel
46. parallel
47. volume
48. mass
49. A. Rigidity
    B. Precession
50. atmosphere
51. vacuum

ANSWERS

52. inches of mercury
53. A. Pitch axis
    B. Roll axis
54. precession
55. deviation
56. followup
57. flight director
58. annunciator

CHAPTER 13

1. does not
2. Federal Communications Commission
3. electromagnetic
4. antenna
5. light
6. electrons
7. 108.0 and 135.95
8. oscillator
9. amplifier
10. modulator
11. A. Amplitude modulation
    B. Frequency modulation
12. crystal
13. demodulator
14. discriminator
15. inverter
16. transceiver
17. do not
18. Restricted Radiotelephone Operator
19. station
20. HF
21. Omnidirectional Range (VOR)
22. course deviation
23. radial
24. omni-bearing selector
25. VOT test
26. red flag
27. A. Localizer receiver
    B. Glide slope receiver
    C. Marker beacon receiver
28. localizer
29. UHF
30. horizontal
31. 75
32. three
33. UHF
34. bottom
35. LF
36. A. Loop antenna
    B. Sense antenna
37. radar beacon transponder
38. do not
39. inertial navigation system
40. radar
41. ground
42. A. 121.5 MHz
    B. 243.0 MHz
43. ELT case
44. bonding
45. shielded
46. static dischargers
47. doubler
48. coaxial

CHAPTER 14

1. hypoxia
2. headache
3. A. Nitrogen

ANSWERS

    B. Oxygen

4. A. 14.7

    B. 29.92

    C. 1013.2

5. tropopause
6. troposphere
7. stratosphere
8. is not
9. cabin outflow
10. structural strength
11. absolute
12. adiabatic
13. engine compressor
14. A. Roots-type compressor

    B. Centrifugal compressor

15. A. Isobaric mode

    B. Constant-differential mode

16. isobaric
17. constant-differential
18. cabin air pressure safety
19. fuel tanks
20. fuel solenoid valve
21. A. Air-cycle cooling system

    B. Vapor-cycle cooling system

22. expansion
23. compressor bleed
24. water separator
25. evaporator
26. condenser
27. condenser
28. evaporator
29. compressor
30. expansion valve
31. 12
32. A. Gaseous oxygen systems

    B. Solid-state oxygen systems

33. is not
34. burned
35. inhales
36. cabin air
37. pressure
38. Castile

asa has more to offer...

For 40 years ASA has been concerned with students. Our goal has always been, and continues to be, providing the most current comprehensive and effective training material possible. As much care goes into our publications as we invest in our teaching. Here's the 1988 line-up of ASA's aviation training material — what we believe to be the BEST available in learning programs and test preparation books for students, whether beginning or advanced.

CASSETTE COURSES

Now you can learn everything you need to know at your own pace. In your home, plane or car — not just to pass the exam, but to give you a thorough working knowledge of the material. Courses are recently-revised, reworked, and include text, tapes and graphics - all in an attractive binder. Buy a course with confidence, knowing you've got the best there is.

PRIVATE PILOT CASSETTE COURSE
Provides basic aviation information, exhaustively, effectively. 200+ pages of text, plus six 90-minute cassettes.
#ASA-CST-P   ISBN 0-940732-12-2

COMMERCIAL PILOT CASSETTE COURSE
Includes a previously untested section on meteorology, and more. 200 pages of text plus six 90-minute cassettes.
#ASA-CST-C   ISBN 0-940732-14-9

INSTRUMENT PILOT CASSETTE COURSE
Rated #1 nationally by an independent group. Easy to comprehend text plus six 90-minute cassettes. Prepares for Instrument and Inst. Instructor exams.
#ASA-CST-I   ISBN 0-940732-15-7

AIRLINE TRANSPORT PILOT CASSETTE COURSE
231 pages of text plus six 90-minute cassettes. Updated, for Part 135 exam.
#ASA-CST-ATP   ISBN 0-940732-19-X

FLIGHT INSTRUCTOR CASSETTE COURSE
202 pages of text plus six 90-minute cassettes. Complete program for both CFI exams — Flight Instructor Airplane & Fundamentals of Instructing.
#ASA-CST-CFI   ISBN 0-940732-18-1

FLIGHT ENGINEER CASSETTE COURSE
Teaches Basic and Turbojet (727). Six cassettes with 230 pages of text. Reference book for FE Test Guide.
#ASA-CST-FE   ISBN 0-940732-16-5

FAA QUESTION BOOKS

All Question Books are exact duplicates of exam questions.

PRIVATE PILOT
T-8080-1A exam. 778 questions with references.
#ASA-8080-1A

COMMERCIAL PILOT
T-8080-2A exam. 1,200 questions and references.
#ASA-8080-2A

FLIGHT INSTRUCTOR
T-8080-3A exam. 507 questions with references, all CFI ratings.
#ASA-8080-3A

FUNDAMENTALS / GROUND INSTRUCTOR
T-8080-4A exam. 792 questions with references.
#ASA-8080-4A

AIRLINE TRANSPORT PILOT (121)
T-8080-5A. For the Airline Transport Pilot Dispatcher test. With references.
#ASA-8080-5A

INSTRUMENT RATING
T-8080-7A exam. 908 questions, references. Includes CFI-I.
#ASA-8080-7A

FLIGHT ENGINEER
T-8080-8A exam. For the Flight Engineer Test. With references.
#ASA-8080-8A

THE COMPLETE PILOT VIDEOS

ASA introduces two of the finest home video courses ever, for the private and instrument pilot!

Produced by a professional TV production team, using striking graphics, and narrated by a meteorologist/TV announcer. These state-of-the-art video cassettes are what you have come to expect from ASA — the finest. The 8-hour cassette courses are bound in an attractive binder — In VHS or Beta format.

THE COMPLETE PRIVATE PILOT VIDEO
A natural extension for ASA! Adds another concept to the learning process, making it "the" way to study at home. We, at ASA, see video as a great educational tool. ASA has set the standard of quality.
VHS — #ASA-VC-P
Beta — #ASA-VC-P/B

THE COMPLETE INSTRUMENT PILOT VIDEO
This video takes off where the Private Pilot Video landed. Four two hour tapes of the most comprehensive Instrument Flight Instruction on the market. Also includes THE COMPLETE INSTRUMENT PILOT KIT.
VHS — #ASA-VC-I
Beta — #ASA-VC-I/B

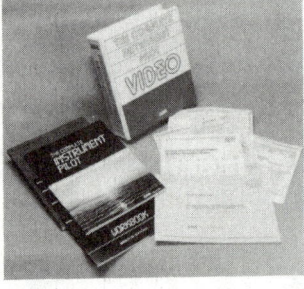

TEST BOOKS

Exact duplicate of FAA written exam questions, includes answers and explanations. Prepared by our specialized research team. Explanation section expanded. An excellent way to go if you want 100% on your next FAA exam.

PRIVATE PILOT TEST BOOK
Reproduction of FAA exam T-8080-1A. 917 questions, with answers and expanded explanations.
#ASA-P010B   ISBN 0-940732-27-0

INSTRUMENT RATING TEST BOOK
T-8080-7A exam. 908 questions with expanded answers and explanations.
#ASA-I040B   ISBN 0-940732-29-7

COMMERCIAL PILOT TEST BOOK
T-8080-2A exam. 1,200 questions with answers and expanded explanations. Includes questions for commercial pilot and the military competency exam.
#ASA-C020B   ISBN 0-940732-28-9

FLIGHT INSTRUCTOR WRITTEN TEST GUIDE
T-8080-3A exam. 507 questions with answers and explanations.
#ASA-CFIB   ISBN 0-940732-30-0

FLIGHT ENGINEER TEST GUIDE
T-8080-8A exam. Referenced to ASA FE Cassette Course. Includes basic, turbojet, and reciprocating engine powered aircraft.
#ASA-FE06B   ISBN 0-940732-31-9

FUNDAMENTALS OF INSTRUCTING / BASIC GROUND INSTRUCTOR TEST GUIDE
T-8080-4A exam. Required exam for all Instructors. 792 questions with answers and explanations. Includes FOI, BGI and Advanced Ground Instructor.
#ASA-FOI-BGI-B   ISBN 0-940732-34-3

AIRLINE TRANSPORT PILOT/ AIRPLANE TEST — Part 121
T-8080-5A exam. 1,372 questions with answers and explanations for Part 121/Dispatcher applicant.
#ASA-ATP121A   ISBN 0-940732-33-5

AIRLINE TRANSPORT PILOT TEST BOOK — Part 135
T-8080-6B exam. 1,053 questions with answers and expanded explanations.
#ASA-ATP-135B   ISBN 0-940732-33-5

AVIATION MECHANIC: GENERAL TEST GUIDE
T-8080-10A exam. 551 questions with answers and explanations.
#ASA-AMG-B   ISBN 0-940732-35-1

AVIATION MECHANIC: POWERPLANT TEST GUIDE
T-8080-11A exam. Contains questions, answers and explanations. For Sections One and Two of exam.
#ASA-AMP-B   ISBN 0-940732-36-X

AVIATION MECHANIC: AIRFRAME TEST GUIDE
T-8080-12A exam. Contains questions, answers and explanations for Sections One and Two.
#ASA-AMA-B   ISBN 0-940732-37-8

TEXT BOOKS

THE COMPLETE PRIVATE PILOT
The definitive textbook for the 80's. Fills a long-standing gap in basic aviation education. Provides student with thorough conceptual and practical understanding, assuring exam preparedness. An EXCELLENT textbook.
#ASA-PPT   ISBN 0-940732-39-4

THE COMPLETE PRIVATE PILOT DELUXE KIT
Kit contains the elements needed by students to effectively prepare for their written test.
#ASA-PPT-KT

THE COMPLETE PRIVATE PILOT WORKBOOK
For use with #ASA-PPT.
#ASA-PPT/W

THE COMPLETE PRIVATE PILOT SYLLABUS
For use with the #ASA-PPT.
#ASA-PPT/S

THE COMPLETE INSTRUMENT PILOT
The newest text from Bob Gardner. Instructs the student using a logical process covering the latest IFR Procedures and Equipment.
#ASA-IPT   ISBN 0-940732-51-3

THE COMPLETE INSTRUMENT PILOT DELUXE KIT
Contains The Complete Instrument Pilot Text, Workbook, T-8080-7A Question Book with answers, Flight Planner and Graduation Certificate.
#ASA-IPT/KT

THE COMPLETE INSTRUMENT PILOT WORKBOOK
For use with the #ASA-IPT.
#ASA-IPT/W

ROTARY WING FLIGHT
Reprint of U.S. Army FM 1-51. For beginning helicopter applicants.
#ASA-RW-4   ISBN 0-940732-38-6

FLIGHT ENGINEER TEST PREP PROGRAM
For both FE Basic and Turbojet (727) exam for FE applicant. Includes Weight and Balance computations.
#ASA-TX-3   ISBN 0-940732-26-2

AIRLINE TRANSPORT PILOT PART 121 TEST PREP PROGRAM
Air Carrier and Dispatcher. Includes Part 121 Regs, Enroute Procedures, Approach Procedures, etc.
#ASA-TX-5   ISBN 0-940732-40-8

JIFFYHOOD
The "new-old" standby for instrument flight check ride. Comfortable, light, yet does the job. Rolls up in your pocket AND reasonably priced. #ASA-H20